PARTICIPANTS

John R. Anderson, O. D. Consultant, The Proctor & Gamble Co., Cincinnati, Ohio

Lorna Arthur, Director, Vocational Rehabilitation, Sound View-Throgs Neck Community Mental Health Center, Bronx, New York

Fern E. Asma, M.D., Assistant Medical Director, Illinois Bell Telephone Company, Chicago, Illinois

Gilbert J. Black, Associate Director, Cage Teen Center, Inc., White Plains, New York

Irving Blumberg, Consultant, International Committee Against Mental Illness, New York, New York

Caesar Briefer, M.D., Physician, Polaroid Corporation, Medical Department, Cambridge, Massachusetts

Robert S. Carson, M.D., Psychiatric Consultant, New York Telephone Company, White Plains, New York

A. Mort Casson, Ph.D., Eagleville Hospital and Rehabilitation Center, Eagleville, Pennsylvania

Stefan J. Cenkner, M.D., Veterans Administration Hospital, Montrose, New York

Sidney Cobb, M.D., Program Director, Survey Research Center, Institute for Social Research, University of Michigan, Ann Arbor, Michigan

Alexander Cohen, Research Psychologist, Behavioral and Motivational Factors Branch, National Institute for Occupational Safety and Health, Cincinnati, Ohio

James L. Craig, M.D., Director of Medical Services, Tennessee Valley Authority, Chattanooga, Tennessee

John M. Daly, M.D., Physician, Standard Oil Company of New Jersey, New York, New York

Arthur Danart, Sociologist, Westinghouse Management Services, Columbia, Maryland

Marjorie T. Dreyfus, Psychiatric Social Worker, Yale University, New Haven, Connecticut

James Driscoll, New York State School of Industrial and Labor Relations, Cornell University, Ithaca, New York

John C. Duffy, M.D., Medical Director, IBM Corporation, Armonk, New York

Roland F. Eller, Chief of Manual Arts Therapy, V. A. Hospital, Montrose, New York

David Fedders, M.D., Psychiatrist, University of Cincinnati, Covington, Kentucky

John R. P. French, Ph.D., Program Director, Research Center for Group Dynamics, Institute for Social Research, University of Michigan, Ann Arbor, Michigan

Walter Gadlin, Ph.D., Psychologist, Sound View-Throgs Neck Community Mental Health Center; Albert Einstein College of Medicine, New York, New York

Sanford Goldstone, Ph.D., Professor of Psychology in Psychiatry, New York Hospital-Cornell Medical Center, White Plains, New York

Craig Grether, Ph.D., Psychologist, Westinghouse Corporation, Columbia, Maryland

Barrie Sanford Greiff, M.D., Psychiatrist, Harvard Business School, Glass Hall, Boston, Massachusetts

Leopold Gruenfeld, Professor, New York State School of Industrial and Labor Relations, Cornell University, Ithaca, New York

Martin Hamburger, Ph.D., Professor of Education, New York University, New York, New York

Francis Hamilton, M.D., Medical Director, New York Hospital-Cornell Medical Center, Westchester Division, White Plains, New York

R. J. Haskell, Jr., Ph.D., Psychologist, IBM Corporation, Hopewell Junction, New York

Howard Hess, M.D., Psychiatrist, Western Electric, New York, New York

R. Edward Huffman, M.D., Psychiatrist, Highland Hospital Division, Duke University Medical Center, Asheville, North Carolina

Eugene T. Hupalowsky, M.D., Psychiatric Consultant, IBM, Corporation, New York, New York

Helen Hurewitz, C.S.W., Director, Mental Health Clinic, South Nassau Communities Hospital, Rockville Centre, New York

Robert Kahn, Ph.D., Director, Survey Research Center, Institute for Social Research, University of Michigan, Ann Arbor, Michigan

George L. Kauer, Jr., M.D., Associate Director, Clinical Medicine, American Can Company, Greenwich, Connecticut

William Kroes, Ph.D., National Institute of Occupational Safety and Health, Cincinnati, Ohio

Harry F. Krueckeberg, Ph.D., Consultant, Indiana Department of Mental Health; Research Director, Bureau of Business Research, Indiana State University, Terre Haute, Indiana

Paul Kurzman, Columbia University School of Social Work, Industrial Social Welfare Center, New York, New York

Thelma Larsh, Educator, C. W. Post College, Locust Valley, New York

Cavin P. Leeman, M.D., Psychiatrist, Beth Israel Hospital, Boston, Massachusetts

Arlene Leonowicz, R.N., Combustion Engineering, Inc., Windsor, Connecticut

Lennart Levi, M.D., Director, Laboratory for Clinical Stress Research, Karolinska Sjukhuset, Stockholm, Sweden

Harry Levinson, Ph.D., President, Levinson Institute; Lecturer, Harvard Medical School; Adjunct Professor, College of Business Administration, Boston University, Cambridge, Massachusetts

William Lhamon, M.D., Chairman, Department of Psychiatry, Cornell University Medical College, White Plains, New York

William T. Liggett, Sociologist, Westinghouse Management Services, Columbia, Maryland

William D. Longaker, M.D., Psychiatric Consultant, IBM Corporation, Endicott, New York

Bruce K. Margolis, Ph.D., Chief, Research and Application Section, National Institute of Occupational Safety and Health, Cincinnati, Ohio

John E. Massey, Director of Personnel, Tennessee Valley Authority, Knoxville, Tennessee

Jermyn F. McCahan, M.D., Medical Director, Long Lines, American Telephone and Telegraph Company, New York, New York

Blair W. McDonald, Jr., Research Psychologist, Navy Medical Neuropsychiatric Research Unit, San Diego, California

W. Edward McGough, M.D., Psychiatrist; Associate Dean, Rutgers Medical School; Consultant, Bell Laboratories, Colts Neck, New Jersey

William McKnight, M.D., Assistant Professor of Psychiatry, New York Hospital-Cornell Medical Center, Westchester Division, White Plains, New York

Alan McLean, M.D., Clinical Associate Professor of Psychiatry and Head, Center for Occupational Mental Health, New York Hospital-Cornell Medical Center, Westchester Division, White Plains, New York

Stanley T. Michael, M.D., Associate Professor of Psychiatry, New York Hospital-Cornell Medical Center, Westchester Division, White Plains, New York

Leo Miller, Manager of Counseling, Polaroid Corporation, Cambridge, Massachusetts

Robert Miller, New York State School of Industrial and Labor Relations, Cornell University, Ithaca, New York

Thomas M. Miller, Commissioner, Industrial Commission of Virginia, Richmond, Virginia

Doris A. Minervini, Instructor, Bronx Community College, Yonkers, New York

Kerry Monick, M.D., Psychiatrist, Mount Sinai Hospital, New York, New York

Eileen Morley, Ph.D., Counseling Psychologist, Harvard Business School, Boston, Massachusetts

Charles R. Pfenning, President, American Train Dispatchers Association, Chicago, Illinois

Betty-Jane Phillips, M.D., Industrial Medicine, Lederle Laboratories, Pearl River, New York

Leon L. Rackow, M.D., Director, Veterans Administration Hospital, Montrose, New York

David L. Ravella, M.D., Psychiatrist, Aluminum Company of America, Pittsburgh, Pennsylvania

P. G. Rentos, Ph.D., National Institute of Occupational Safety and Health, Cincinnati, Ohio

David B. Robbins, M.D., Psychiatric Consulant, IBM Corporation, Harrison, New York

Robert M. Rose, M.D., Psychiatrist, Boston University Medical Center, Boston, Massachusetts

W. Donald Ross, M.D., Professor of Psychiatry, University of Cincinnati, Cincinnati General Hospital, Cincinnati, Ohio

Herbert L. Rothman, M.D., Psychiatric Consulant, IBM Corporation, New York, New York

David R. Schwarz, Management Consultant, Larchmont, New York

Edward E. Seelye, M.D., Assistant Professor of Psychiatry, New York Hospital-Cornell Medical Center, Westchester Division, White Plains, New York

Charles A. Shamoian, M.D., Ph.D., Psychiatrist, Payne Whitney Psychiatric Clinic, Cornell Medical Center, New York, New York

Arthur B. Shostak, Ph.D., Associate Professor, Drexel University, Department of Social Sciences, Philadelphia, Pennsylvania

John Sommer, M.S.W., Industrial Social Welfare Center, Columbia University School of Social Work, New York, New York

Graham C. Taylor, M.D., Psychiatrist, McGill University, Montreal, Quebec

Granville I. Walker, Jr., M.D., Medical Director, Chase Manhattan Bank, New York, New York

Porter Warren, M.D., Assistant Medical Director, New York Hospital-Cornell Medical Center, Westchester Division, White Plains, New York

Clinton G. Weiman, M.D., Medical Director, First National City Bank, New York, New York

Louis Jolyon West, M.D., Professor and Chairman, Department of Psychiatry, School of Medicine, University of California, Los Angeles, California

Walter L. Wilkins, Psychologist, Navy Medical Neuropsychiatric Research Unit, San Diego, California

PREFACE

To CAPTURE the best of the two days' discussion in several arenas from professionals in different disciplines is one challenge this volume attempts. Much came before and a great deal followed the seminal conference which also appear in these pages.

There were seven major presentations; each represented here in one way or another. In two instances the presentations are verbatim. Doubling as conference reporter I summarized two. The final three represent the authors' later elaborations.

There were six subgroup discussions. In one instance the intensive interaction of such a group led both to an innovative recommendation for greater interplay among key executives and occupational physicians and to a chapter in this volume. Some discussions yielded stimulating interchange but did not produce publishable material. Others are commented on herein. There is also a chapter of background material and an interpretative summary by the editor.

The format in this volume has been altered from the conference protocol to present a cohesive whole, to highlight the productivity of those who so successfully worked with us and who deserve the major order of credit for this contribution.

At the Center we are not only grateful to all participants, particularly those who served as speakers and discussion leaders, but also to those who helped fund our work: The National Institute of Occupational Safety and Health*, The U. S. Steel Foundation, and the New York Hospital-Cornell Medical Center, Westchester Division.†

Our concern with occupational stress at the Center for Occupational Mental Health grows largely from prior efforts to use this format of interdisciplinary study groups to assemble monographs and

---

*Grant No. 1 R13 OH-00429-01

†For permission to reprint Chapters 2, 5, 6, 7, 8, 10, 11 and 12 we are grateful to *Occupational Mental Health,* copyright by the Center for Occupational Mental Health of the Department of Psychiatry of Cornell University Medical Center.

papers which represent a distillate of the state of the art for a particular readership.

After becoming established as an information center concerned with developments relating to the adaptation of the individual to his work, Center staff created abstracting and dissemination capability under contract with the National Clearinghouse for Mental Health Information. As we observed the vast amount of published information and the number of programs concerned with fostering mentally healthy behavior and coping with the disturbed behavior of those who are ill at work, it became clear that the need was great to bring together professionals from disparate fields whose areas of concern were similar, to exchange views in the hope that each participant might be encouraged to broaden his outlook. We further saw a continuing role for the Center in translating and interpreting data from the clinical and behavioral sciences for those in positions of authority in work organizations and for professionals in occupational medicine and nursing.

The latter concern led to cosponsorship in 1966 with the National Association of Manufacturers of a conference on Mental Health and the Business Community. The symposium that resulted was seen to be of concern to all levels of business management. Nineteen qualified professional participants led discussions concerning a wide range of related subjects. It tackled such questions as: What is mental health anyway? How does it relate to the work we do? Does it relate to productivity, to job satisfaction? Is it possible for an employer to be other than economically selfish in his concern for the mental health of those in his employ? What should an employer do when an employee develops a mental disorder? What is an employer's legal liability for claims relating to work-induced mental disability? Are top management people more prone to hypertension, ulcers, heart disease and psychosomatic disorder? And, finally, isn't there enough of big brother in the world to make unnecessary this sophisticated involvement of psychiatrists, psychologists and industrial physicians employed by the company?

These matters were discussed for the 300 participants who registered for a one-day symposium at the Waldorf Astoria in March of that year. A more detailed and widely seen interpretation of the conference was published in the fall of 1967*.

From 1967 to 1969 the Center concentrated on the first detailed interdisciplinary study of current thinking on several topics relating to persons in work organizations. There were frequent meetings and many position papers which discussed theoretical considerations, programs and philosophy. The results, designed for a professional audience, were published in 1970†. A nucleus of participants has continued to meet as a study group and as a core of professionals who strongly influence the direction of Center projects.

Occupational Mental Health

The Center's regular publication, *Occupational Mental Health,* continues to carry abstracts of relevant literature, original articles and commentary as well as news of activities in the field. Since 1970 this quarterly has been distributed to some 4,000 subscribers essentially as a successor to *Occupational Mental Health Notes* which for several years had been published by the National Institute of Mental Health from material provided by the Center and distributed by the U. S. Government Printing Office. Both publications have, until the time of this writing, been published gratis through the generosity of agencies such as the Ittleson Family Foundation and sponsors of the Center's various activities. The contents of *OMH,* representing as it does a cross section of activity in the various disciplines concerned with the field, continued to suggest the need to focus on occupational stress. Thus, the next major step in the growth of the Center was the conference reported in this volume.

Occupational stress has been studied and defined from many frames of reference by researchers and clinicians in several disciplines. The purpose of the 1972 Occupational Mental Health Conference was to bring together representatives of the many concerned disciplines to present their points of view so that a picture, or at least part, of the current thinking on the subject could be discussed. Drawing

---

*McLean, A. A.: *To Work Is Human: Mental Health and the Business Community.* New York, Macmillan, 1967.

†McLean, A. A.: *Mental Health and Work Organizations.* Chicago, Rand, 1970.

from the formal presentation, several small group discussions centered on definitions of occupational stress. Some of their conclusions were sufficiently cohesive to include in this interpretation of the conference.

White Plains, New York   ALAN McLEAN, M.D.

CONTENTS

OCCUPATIONAL
STRESS

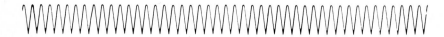

CHAPTER 1

CONCEPTS OF
OCCUPATIONAL STRESS, A REVIEW

ALAN McLEAN

A BRIEF SUMMARY of examples of approach to occupational stress
will serve as an introductory frame of reference to concepts
elaborated upon in this volume. My intent is to introduce and
illustrate a few of the current frames of reference; not to document
all or even most points of view. I particularly acknowledge omission
of much that is classic such as the early experiments demonstrating
relationships between emotional status of individuals and their somatic
reactions. Cannon's proposals of fight-flight reactions serve as one
example (1929). The studies of Wolff (1950) and Wolf (1950)
at Cornell also cannot be ignored in the history of stress response.
Nor do I suggest doing so. Indeed, presentations in this volume
owe allegiance to these foundations.

PHYSIOLOGICAL METHODS OF MEASURING
OCCUPATIONAL STRESS

Persistent or recurring emotional stress has been considered an
important etiological factor in a number of physical and psychiatric
disorders. Several investigators have developed objective measures
to quantify such reaction to stress. The work of Dr. Lennart Levi,
one of the conference speakers, exemplifies a variety of physiological
methods used to measure industrial stressors.

His research team has exposed groups of subjects to a variety of
stimuli, including laboratory as well as real life stresses. They have
used such laboratory situations as simulated industrial work; simulated
office work; films chosen to induce anxiety, aggressiveness and other

emotional reactions; simulated flight with simultaneous control of an aircraft and a guided missile; and prolonged function under simulated ground combat conditions. They have studied actual work situations including those of telephone operators and invoicing clerks. The influences of differing noise levels and of office landscaping have also been researched.

Responses to stressful situations have been assessed with respect to psychological self-ratings, performance, and biochemical parameters reflecting the subjectively perceived stress in the organism. In a series of such studies Dr. Levi has clearly demonstrated that psychological stimuli can actually affect physiological and biochemical reactions in a potentially pathogenic manner.

In one experiment lasting seventy-five hours, thirty-two senior military officers alternated between three-hour sessions on an electronic shooting range and at military staff work. Such a regimen of both psychomotor and intellectual tasks is present in many civilian occupational settings as well. No relaxation or sleep was allowed, nor were stimulants, smoking or walking. Although the emotional reactions thus provoked were of moderate intensity, significant biochemical changes were found to occur at the end of the test period. The erythrocyte sedimentation rate increased by 38 percent. Protein-bound iodine increased by 19 percent. Serum iron decreased by 26 percent.

It has been hypothesized that a correlation exists between occupational stress and coronary heart disease. This was the starting point for a second study carried out by Dr. Levi in collaboration with Drs. Lars A. Carlson and Lars Fuller. It was based on several investigations indicating that chronic exposure to emotional stress leads to increased plasma concentration of cholesterol and very low density of lipoproteins. A chronically elevated blood lipid content combined with a direct cardio-toxic effect of the adrenal hormones has in turn been considered capable of contributing to the degenerative changes in the myocardial tissue. Therefore, a second object was to block specifically the stress-induced release of these fatty acids and, consequently, the subsequent increase in plasma lipids by administering nicotinic acid.

Thirty-three male volunteers were divided into three matched groups of eleven each. One group serving as a control was allowed to sit calmly and comfortably throughout three 2-hour periods.

The second group was exposed during the second of the three 2-hour periods to a situation involving sorting ball bearings to the accompaniment of distracting criticism, noise and light, thus imitating what seems to be some of the important stresses of industrial and craft work. The routine was the same for the subjects of the third group as for those in the second group except that they were given a total of three grams of nicotinic acid during the first half of the experiment.

According to Levi's hypothesis, nicotinic acid treatment should block the stress-induced mobilization of the fatty acids. This blockage should take place locally in adipose tissue without affecting the simultaneous hemodynamic and catecholamine responses. Their results support this hypothesis. Thus the heart-rate increase was significant but similar in both stressor-exposed groups and absent in the control group. Corresponding results were found with respect to systolic blood pressure and excretion of adrenaline and noradrenaline. Stress-induced increases in free fatty acids occurred in accordance with their hypothesis. Furthermore, this increase in the free fatty acids was significantly blocked by the nicotinic acid treatment, as was an increase in plasma triglycerides, which actually showed a significant decrease.

This study demonstrated that increased plasma triglyceride levels, probably in response to an increased sympatho-adrenomedulary activity, are readily induced even by a work situation which is not real, but simulated, of short duration and moderate intensity.

In a non-laboratory setting Levi's group studied twelve young, healthy female invoicing clerks during four consecutive days when performing their usual work in their usual environment. The primary objective of this study was to map out the change, if any, in productiveness, subjective feelings, and urinary catecholamine excretions when the usual mode of remuneration of the subjects was changed from salary to piecework wages.

The second and the fourth days of the experiment were control days when the subjects worked for the modest monthly salary. On the first and the third experimental days, piecework wages were added to the subjects' salaries. During these days their income accelerated with their productiveness up to an attainable maximum of 400 invoices per hour, yielding an extra income of more than $80.00 for two days' work. Nothing in the experimental setting was changed except the mode of remuneration. The results may be summarized as follows: During the salary control days, the output of work was very close to the previously calculated ordinary productivity. On piece-work days, the output rose significantly—by no less than 114 percent. However, this very high output of work was accomplished at the expense of considerable feelings of mental and physical discomfort. Half the group complained of feeling hurried and all but two complained of fatigue, backache, and pain of the shoulders or arms during the piecework days. During salary days physical complaints were virtually absent. During salary days the mean urinary excretion of adrenalin and noradrenalin was 5.5 and 18.5 nanograms per minute respectively. During piecework, the mean adrenaline and noradrenaline excretion rose by 40 percent and 27 percent respectively, the changes being statistically highly significant.

All the results point to a subjective, as well as an objective, state of stress reactions, not a severe one but nevertheless induced by the piecework conditions of this experiment. (No judgment was intended about the effects of piecework conditions as such.) During salary work only insignificant signs of such a state were recorded. It might be claimed that such a stress state if prolonged could be detrimental to health in susceptible individuals. On the other hand, it is possible that an adaptation may occur to throw further light on these problems.

Dr. Levi elaborates and draws important conclusions from these data in Chapter 5.

There has, of course, been a great deal of work similar to that from the Karolinska group. The history of psychophysiological reaction to stress is well documented and need not prolong this protocol. One interesting area of overlapping research is that with the group from the University of Michigan using role theory as their frame of reference in studies of the relationship between stress and coronary heart disease.

ROLE THEORY

How do persons at work respond to situations which are, to them, stressful? Stress reactions can be physiological, resulting in psychosomatic reactions such as hypertension, peptic ulcer, or cardiac disease. Or stress can result in impaired performance, increased rigidity of behavior, or reduced problem-solving ability. In contrast, performance can be unaffected, or even improved, by certain kinds of stress, especially when appropriate response has been learned previous to the onset of the stress. Under stressful circumstances, perception is often disrupted and somewhat less adaptive than under normal conditions. A person under stress is less likely to tolerate an ambiguous situation; he is apt to seek more information, whether useful or not. Most people try to cope with job stress either passively, by working harder, or actively, by changing circumstances and environment. One possibly overlooked technique of coping with stress such as information overload (excessive and conflicting input about one's expected role) is to make other members of one's role-set more aware of the problem. Organizational citizenship should give everyone the right to bring together those in his role-set with a goal of reaching agreement concerning each member's individual duties and performance.

In a therapeutic mode, one may consider that if a person knows beforehand that a stress situation is developing, he can often take steps to mediate that stress. Fear of consequences is excellent motivation to deal with stress, especially in situations of overload, reorganization, or technological change. But since role conflict or ambiguity are seldom foreseen, prior successful experience or knowing someone with experience, as in a professional relationship, can be most helpful in mediating the effects of the stressors. (McLean, 1970)

JOB STRESS AND HEART DISEASE

Work overload has been related to both excessive smoking and increased levels of cholesterol. (French and Caplan, 1971) These authors noted that both quantitative and qualitative overload correlate with job related threat. The number of cigarettes smoked, a much publicized risk factor of heart disease, also increases with increases in the actual number of phone calls, actual visits and meetings a person has (as tallied by each person's secretary). The correlation between cigarette smoking and this measure of quantitative overload is .58.

Several studies have reported an association of work overload with cholesterol level, but in these, only crude indicators of overload were used such as tax deadline for accountants (Friedman, Rosenman and Carroll, 1958), and an impending examination for medical students (Dreyfuss and Czaczkes, 1959, etc.). Accordingly a preliminary field study by French and Caplan tested the hypothesis that overload increases both cholesterol level and heart rate.

As predicted, both objective and subjective overload were substantially related to heart rate ($r = .39$ and $.65$ respectively). Both were also related to cholesterol ($r = .48$ and $.41$ respectively). These findings are based on correlations across individuals; but analyses within individuals showed that increases in observed workload were accompanied by increases in heart rate, a fact which supports their hypothesis that stress produces the observed physiological strain.

The same authors evaluated role ambiguity as stressful. They defined such ambiguity as a state in which the person has inadequate information to perform his role in an organization. Drawing largely from Kahn *et al.* (1964), they observed that major findings showed that men who suffered from role ambiguity experienced lower job satisfaction and high job-related tensions. They went on to conclude from their own work that, with regard to role ambiguity, it was significantly related to low job satisfaction and to feelings of job related threat to one's mental and physical well-being. In addition they observed that the more ambiguity the person reported, the lower was his utilization of intellectual skills and knowledge, and the lower was his utilization of his administrative and leadership skills. This lack of utilization also adversely affected satisfaction and increased job-related threat. In summary, French and Caplan in several studies show that the various forms of workload produce at least nine different kinds of psychological and physiological strain in the individual. Four of these (job dissatisfaction, elevated cholesterol, elevated heart rate, and smoking) *are* risk factors in heart disease. They conclude that it is reasonable to predict that reducing work overload will reduce heart disease.

This same series of studies led French and Caplan to observe that stress manifests itself in forms which can hardly be considered conducive to good health and low risk of coronary heart disease. First, the more time the individual spends carrying out responsibility for the work of others, the more he smokes. Second, the more responsibility he has for the work of others, the higher is his diastolic blood pressure. On the other hand, the more responsibility he has for objects, the lower is his diastolic blood pressure.

OCCUPATIONAL DIFFERENCES

People with different jobs encounter different types of stress and different quantities of stress. Consequently, they experience different types of strain. French and Caplan's research shows that administrators, for example, are far more likely to develop coronary heart disease both because they experience different types of organizational stress linked to risk factors in the disease and because they experience more of those types of organizational stress than do engineers and scientists.

They offer a prescription based on prevention, commenting that when a man dies or becomes disabled by a heart attack, the organization may be as much to blame as are the man and his family. But the assignment of guilt will not prevent people from dying nor prevent the organization from incurring large losses in terms of the investments required to recruit and train the person and to replace him with someone else. What is needed instead is a program of organizational diagnosis and prevention aimed at curtailing the loss of human resources to an organizational enemy known as coronary heart disease.

Although acknowledging that reviews of the medical literature suggest that the known risk factors in coronary heart disease account for only about 25 percent of the variants of the disease, they suggest that there may be some justification for believing that changing occupational stress may be more effective than other approaches to prevention.

Drs. Cobb, Kahn and French discuss this data, elaborate from their many additional studies of occupational stress and strain, and draw significant new conclusions in Chapters 6, 7 and 8.

The one suggestion derived from their experimental program involving work group participation suggests that when the employee feels supported rather than threatened by his boss he feels under less stress.

Generalizing these findings a bit, we suggest that attempts to decrease stress by increasing participation should also try to provide a supportive supervisor and a cohesive and supportive group of co-workers. Such supportiveness will directly reduce psychological strain, and it will also increase the effectiveness of the participation.

Participation must be meaningful, not illusory. Participation in trivial decisions is essentially meaningless. The basic principle is one which involves the relevance of participation to the individual.

Finally these authors conclude that the decisions in which people participate should be perceived as being legitimately theirs to make.

If a group does not feel it is deciding on something within its area of freedom, the participants will tend to feel anxious, threatened, and even dissatisfied.

The body of evidence from studies emenating from those with an orientation to role theory lends strong support to the notion that modern organizations have an impact on both the psychological and physiological health of their members. Many of the stresses which are fairly prevalent in national samples and in specific organizational settings appear to be linked in one way or another with strains which produce coronary heart disease. But the fact that coronary heart disease seems to be as much a part of organizational life as do other traits of the organization, such as size and structure, does not mean that steps cannot be taken to reduce the risk of disease. There are innovative and pioneering steps which management can take. Programs which involve the coordinated effort of management, medical personnel, organizational psychologists can be developed. Careful evaluation of these programs can be carried out; and through such experimental programs, modern organizations can make potentially important contributions to both the management and medical sciences with benefit for both the individuals and the organization's well-being and strength. (French and Caplan, 1971)

PSYCHODYNAMIC CONCEPTS OF STRESS

Clinicians would agree that the major source of stress reactions lies in feelings which may be difficult to fully assess or understand. They would be inclined to argue that often an individual does not really know *what* he is feeling, and certainly much of the time is unaware of *why* he feels as he does.

Man has many different kinds of emotional reactions and great variation in the intensity with which he experiences them. Like all other animals, in Freudian terms he must manage the basic drives of sex and aggression. Both are necessary for the survival of the species. In man, these drives give rise to feelings of love and hate. A major problem is that the more important a person is to you, the more vulnerable you are to his behavior and the more early you may become angry with him. Since it is quite difficult both consciously and unconsciously to love and hate the same person simultaneously, we tend to repress angry feelings from awareness.

In this frame of reference, stress may be perceived as threatening events. One may think of three major threats to the balance of one's psychological equilibrium. Any one of them may cause an individual to react defensively—with anxiety.

The first is the threat of losing control of oneself. At times an individual literally becomes so angry he could kill someone if he lost control. Ordinarily the reaction is to leave the situation or avoid provocation.

A second major threat is the threat to one's conscience or super-ego. The experience of guilt for having done something wrong is not uncommon. But often feelings of guilt arise merely in thinking about something which one believes to be wrong. At times one distorts the meaning of one's own behavior to define it as wrong which further increases guilt.

The third major threat from a psychoanalytic frame of reference is that of personal physical harm. But fear is not a great problem as far as psychological stress goes because one is usually aware of what one is afraid of and can therefore cope with it. Fear becomes a problem only when it is prolonged and we cannot escape the threat, or when it is irrational.

Therefore one can identify as stressors factors which are productive of fear, the threat to one's superego, and those which may stimulate feelings of losing control of oneself.

Psychological stress may be psychoanalytically defined as the activation of any one of these three threats, and there is a wide variation in each individual's vulnerabilty. While there are events and circumstances which threaten all of us, each individual has what Menninger referred to as "an Achilles heel"—emotional conflicts, the triggering of which by an environmental event can produce overwhelming anxiety.

There is also a common denominator to most occupational stress from the psychodynamic viewpoint—change. All change involves loss of some kind—familiar faces, places, pleasures, ways of doing things, organizational supports, etc. Promotions, demotions and transfers, however desired, are changes. Such losses are more severe than many individuals recognize. Change is a threat to the ways people have developed to handle their dependency needs as well. (Levinson, 1969, 1970).

There has been considerable study of stresses in various occupations—both psychological and physical. Studies of miners, ministers, physicians and executives appear among the most common of those conducted by researchers from a psychodynamic frame of reference. For instance, irregular work and inordinate dependency on company facilities in a coal mining area of Virginia have been suggested by Field, Ewing and Mayne (1957) as determinants of a high rate of psychiatric disorder among miners. Stress above ground was found to be as important as the underground hazards that are commonly believed responsible.

Attention was focused on executive stress by three chapters in the volume, *To Work Is Human,* by Taylor, Warshaw and Wright (1967).

Levinson's volume entitled *Executive Stress* is an assemblage of twenty-four articles on the subject.

SELYE'S CONCEPTS

> Most of the agents which can make us sick are, to a greater or lesser extent, *conditionally acting disease-producers*: that is, they cause maladies only under certain circumstances of sensitization. Their disease-producing ability may depend upon hereditary factors, the portal through which they enter into the body, the previous weakening of resistance by malnutrition or cold, etc. Here it is impossible to decide which, among the many factors necessary for the production of disease, is actually the cause and which the conditioning factor: a complete *disease-producing situation* must be realized before the malady can become manifest . . . The development of illness depends upon a whole constellation of events.

Selye specified the importance of time, space and intensity. "Apart from the relative proportions between these three basic factors of disease, the manifestations of individual maladies would then only depend upon *when, where* and *how much.* This tripartite situation develops: upon (1) *time* (the duration and possible repetitive nature of stress), (2) *space* (the location of the stress situation), and (3) *intensity* (the degree of the whole tripartite phenomenon)."

He continued to develop his reaction theory.

> Although biologic reactions tend to give the impression of oneness, they actually represent mosaics of simple activation and inhibition (stress) in a great variety of preexistent, elementary, subcellular, biological units: the *reactions*. Each of these is capable of only one kind of response, but by blending these elementary reactions in different combinations, qualitatively distinct aggregates result.
>
> One may therefore conceive of physiological, biological or psychological events as occurring, not through the mere summation of distinct elements, but through the function of the (*Gestalt*) shape as a single unit. The shape of each disease functions as a single unit, although it is made up of innumerable simple reaction responses. (Selye, 1956)

SUMMARY

In summary, occupational stress is generally seen and measured by individual reactions to situations perceived consciously or unconsciously as threatening. People react based on personal predisposition and in a vocational context. Such reactions are generally of tremendous complexity. Seen by the clinician as symptoms to be treated and by the researcher as events to be measured, they have other dimensions as well. There are the relationships between stress and productivity, morale and job satisfaction. Legal ramifications include employer responsibility under workmen's compensation statutes, federal standards, and state and local regulations.

BIBLIOGRAPHY

Cannon, W. B.: *Bodily Changes in Pain, Hunger, Fear and Rage: An Account of Recent Researches into the Function of Emotional Excitement.* New York, Appleton, 1929.

Dreyfus, F., and Czaczkes, J. W.: Blood cholesterol and uric acid of healthy medical students under stress of examination. *Archives of International Medicine, 103*:708, 1959.

Field, L. W., Ewing, R. T., and Mayne, D. W.: Observations on the relation of psycho-social factors to psychiatric illness among coal miners. *Int J Soc Psychiatry, 3*:133-145, Autumn, 1957.

French, John R., and Caplan, Robert D.: Organizational stress and individual strain. Unpublished paper, University of Michigan, Institute for Social Research, 1971.

Friedman, M., Rosenman, R. H., and Carroll, V.: Changes in serum cholesterol and blood clotting time in men subjected to cyclic variation of occupational stress. *Circulation, 17*:852-861, 1958.

Kahn, R. L., Wolfe, D. M., Quinn, R. P., Snoek, J. D., and Rosenthal, R. A.: *Organizational Stress: Studies in Role Conflict and Ambiguity.* New York, Wiley, 1964.

Levi, Lennart: Emotional stress and biochemical reactions as modified by psychotropic drugs with particular reference to cardiovascular pathology. Report from the Departments of Internal Medicine and Psychiatry, Karolinska Hospital, Stockholm, Sweden, July 1968, 26 pages.

Levi, Lennart: Physioligical methods of measuring industrial stress. *Occupational Mental Health Notes,* April, 1970, pp. 3-6.

Levi, Lennart: Sympatho-adrenomedullary responses to emotional stimuli: Methodologic physiologic and pathologic considerations. In Bajusz, E. (Ed.): *An Introduction to Clinical Neuroendocrinology,* New York, S. Karger, 1967, pp. 78-105.

Levinson, Harry: *Executive Stress.* New York, Har-Row, 1970.

Levinson, Harry: Managing executive stress. *Panhandle Magazine, 4*:16-19, Winter, 1969-1970.

Levinson, Harry, and Weinbaum, Lewis: The impact of the organization on mental health. In McLean, Alan (Ed.): *Mental Health and Work Organizations.* Chicago, Rand, 1970.

McLean, Alan: *Mental Health and Work Organizations.* Chicago, Rand, 1970.

McLean, Alan: *To Work is Human: Mental Health and the Business Community.* New York, Macmillan, 1967.

Selye, Haug: *The Stress of Life.* New York, McGraw, 1956.

Wolff, H. G.: Life stress and bodily disease—a formulation. In *Life Stress and Bodily Disease.* Volume 29, Chapter 69, *Association for Research in Nervous and Mental Diseases, Proceedings,* Williams & Wilkins, Baltimore, 1950.

Wolff, H.G., Wolf, S., and Hare, L.C.: *Life Stress and Bodily Disease,* Vol. 29, *Association for Research in Nervous and Mental Diseases, Proceedings,* Williams & Wilkins, Baltimore, 1950.

CHAPTER 2

OCCUPATIONAL STRESS AND STRAIN

Bruce K. Margolis, William H. Kroes

JOB STRESS IMPLIES so many events and processes that it is often conceived to be a nebulous concept, difficult to study in a scientific manner. But, there is at least one meaningful paradigm from which significant research is readily derived. Job stress may be defined as the condition in which some factor, or combination of factors, at work interacts with the worker to disrupt his psychological or physiological homeostasis. The factor or combined factors at work are generally called job stressors, and the disrupted homeostasis is often called job-related strain. The concept of factors *interacting* with the worker is significant. It is quite clear from informal observation, as well as controlled research, that individuals respond to identical job situations in very different ways. One worker is upset by a boss who closely supervises his work while another finds close supervision desirable. It is for this reason that French and others at the Institute for Social Research, University of Michigan conceive of job stress as a poor *person-environment* fit. When the worker's needs are frustrated or his abilities mismatched with his responsibility, job-related strain is likely to result.

In order to better understand the impact of job stress we must be able to measure its effects on the worker. Too often, job-related strain is conceived of as being only unidimensional, or at most bidimensional. However, at least five dimensions of job-related strain can and should be measured in order to appreciate the effects of job stress upon the worker.

First are the subjective states of anxiety, tension, anger, feeling uptight and the like. These are short-term, rather than chronic states, which occur in close temporal proximity to specific job stressors.

[15]

The boss bawls you out, creating a temporary state of anger, fear or anxiety, but these feelings are no longer present when the boss compliments you for a good job the following week.

This kind of response can be differentiated from a second dimension of job-related strain, the more chronic psychological responses to job stress. Chronic depression, feelings of fatigue, alienation, or general malaise become part of the worker's health status rather than a reaction to specific circumscribed work situations. Just as hypertension is distinguished from acute blood pressure, so can chronic mental health problems be distinguished from acute psychological responses to job stress.

The third measure of job-related strain involves transient clinical-physiological changes. Levels of catecholamines, blood lipids, blood pressure, gut motility, and other physiological variables change significantly under psychological stress and serve to indicate in an objective way that the individual is under stress.

The fourth measure is physical health status. Such consequences of job stress as gastro-intestinal disorders, coronary heart disease, asthmatic attacks, and other psychosomatic disorders are all established as consequences of psychological stress. More and more, reports of physical health consequences of job stress can be seen in magazines and newspaper articles.

The final measure of job-related strain is work performance decrement. Laboratory research and some field research have, for years, been demonstrating severe decreases in productivity and increases in error rates as a function of psychological stress associated with the task.

Having enumerated five measures of job strain, let us turn to the goals of job stress research. Clearly they depend very much on who is sponsoring or conducting the study. For industry a primary goal might be to understand the causes of performance decrement so that productivity can be improved. For the National Institute for Occupational Safety and Health (NIOSH) the health consequences of job stress, both mental and physical, are of central concern. But in order to fully understand the health consequences, an understanding of the subjective and physiological reactions is also required. In fact, for industry to understand the causes of work decrement they too must understand the subjective and physiological reactions to job stress as well as their health consequences, as these five indices of job-related strain are not discrete, but all derive from the same source. They are inter-dependent, with subjective state perhaps causally related to mental health status, physiologic state causally related to physical health status, and all four of these causally related to performance decrement.

It is certainly true that NIOSH has a responsibility to study job stress as it affects health and safety, but it is equally true that industry has a vested interest in job stress studies. NIOSH's interest derives from the mandate of the Occupational Safety and Health Act of 1970 which specifically requires tht NIOSH conduct research into the psychological factors involved in worker safety and health. Industry's interest should derive from a desire to maintain a healthy and productive work force.

Now that we have emphasized the importance of research on job stress, let us examine some intrinsic methodological problems which every field or *in situ* study must face. The basic goal in job stress studies is to relate specific job stressors to specific strains so that those stressors which are causing problems might be dealt with. One problem lies in being able to draw conclusions when high correlations between presence of stressor and presence of strain are found. There are so many variables which could produce such correlations independent of job stress effects. For example, factory workers on the assembly line may have higher rates of depression than their bosses, not because of their work but because of their poor quality home life. Physical hazards, such as toxic agents on a factory floor, may cause asthmatic attacks which are incorrectly attributed to the psychological stresses of the job. Racial prejudice, felt or real, can be a source of stress for the worker, independent of the requirements or nature of the job.

On the other hand, it is also possible that job stress might be present even when scant correlations between stress and strain are found. In those cases where individuals susceptible to particular job stressors are removed from the job through self-selection or weeding out on the part of management, the remaining workers will in general show very little job-related strain. Understanding of the job stress problem even when strain is not evident, however, might make for better prior selection of workers, or modification of the stressors so that the worker pool can be expanded.

Another kind of problem involves the validity of separating job stressors and treating them as additive. Typically we look at jobs high on one kind of stressor but we don't consider the synergistic, or possible ameliorating effects of different arrays of job stressors. Stressors can be studied independently, but the whole work situation must be examined to draw valid conclusions about the job's impact on the worker.

In our own research program we have attempted to take account of these methodological factors. Probably our most significant effort involves the evaluations of job stressors and their consequences for about 1,000 workers in twenty categories. All but a few jobs have substantial job stress from one or more stressor sources. On a contractual basis and under the direction of J.R.P. French and Sydney Cobb of the Institute for Social Research a carefully designed study has been laid out. In broad outline, the following steps will be taken:

1. High stress jobs and the specific stressors therein will be identified based upon judgments of individuals highly familiar with those jobs; a few lower stress jobs will also be identified for comparison purposes.

2. Individual workers will, via interview and questionnaire, describe their perceptions of the stresses under which they work and their subjective reactions to them.

3. Worker medical records will be evaluated where they are both available and useful; in addition, medical histories will be taken for each worker in the study.

4. Clinical-physiological tests evaluating catacholamine levels, blood pressure, and other appropriate physiological indicators of stress will be made on a subsample of 300 workers.

5. Statistical analyses will be performed relating specific jobs and specific job stressors to strain outcomes.

Of the five strain measures, only work performance decrement will be omitted from this study. Here companies participating in the study can make their own work performance measures and inexpensively gather information relevant to productivity problems.

The payoff of the study can be substantial from the point of view of understanding the impact of job on worker health and enabling us to focus upon those jobs and those stressors which are most significant. From this we hope to develop remediation programs based upon a logical and empirically derived set of priorities.

Coordinated with this study, NIOSH will be piggy-backing onto a triennial interview survey of a statistical sample of 1500 American workers. This study's general purpose is to get a reading as to the American worker's reactions to his job. Now, additional questions will be asked in probing for perceived psychological stressors at work and any mental and physical health problems the worker may have. Clearly though, such a survey by itself cannot provide unequivocal support for a cause and effect relationship between job stress and worker health, it is expected to provide further evidence as to the nature of this relationship for a broad range of occupations.

A third study now being conducted by NIOSH in-house is an attempt to identify job stressors for policemen and to determine the health impact of job stress among policemen. A pilot study consisting of interviews with 130 policemen in the city of Cincinnati is now underway. If the preliminary data warrant it, a more detailed and analytical study will be developed and performed on a larger scale.

Another study just beginning involves one particular occupation for which the stress of threat to one's life and limb may be substantial —underground coal mining. The extent to which coping mechanisms are operating to deal with this threat, but at the same time operating to increase risk-taking and accident potential, will be an essential question in this study. Fifteen hundred mine and management personnel as well as union officials and miners' wives will be interviewed to determine the attitudes and feelings of miners toward their work and of management toward mine safety, and the impact of these attitudes and feelings on miner health and safety. Ultimately, we hope to develop techniques of modifying those attitudes, feelings and behaviors which are associated with increased risk of ill health or accidents.

Psychological stress at work is a serious and complex problem affecting millions of workers across the country. The National Institute for Occupational Safety and Health has made a commitment to the goal of adding to our understanding of job stress and finding ways to reduce its adverse impact on worker safety and health. It is hoped that industry will combine with health practitioners and researchers to facilitate achievement of that goal.

CHAPTER 3

CLINICAL CONCEPTS*

A T FIRST GLANCE one would think there are both specific and non-specific factors on the job which are stressful to each job holder. But this is only at first glance. Reactions to stress can only be understood in the context of the job holder's entire life situation. A job is part of life and, as was so clearly demonstrated by Dr. Harold Wolff earlier, it is one's *total* life stress which must always be considered and this is extremely complex.

The involved quantification which marks the research from the University of Michigan reported in Chapters 6, 7 and 8 is indeed interesting and the correlations suggestive. The authors cannot begin however to fully assess the vast array of occupational and non-occupational variables even though the relationships which they bespeak are interesting.

Dr. West, in speaking on this issue, cited an anecdotal example of military experience during the Second World War in which soldiers were being carefully selected for work in the Arctic. A vast array of tests and biographical data were studied and later correlated with successful and unsuccessful adaptation. Only one item was highly significant in its positive correlation with successful adaptation. It was a single question which asked each soldier, "Would you *like* to work in a cold climate?"

The question of accepting relatively simplistic clinical data, such as asking the patient, is often lost sight of in much research that relates to occupational stress. It *never* should be.

The importance of non-occupational variables is suggested in

---

*This chapter is largely based on the lengthy, but informal, presentation of Dr. Louis Jolyon West. The editor prepared this chapter from tape recordings of the conference proceedings.

Chapter 5 with reference to the Holmes-Rahe scale which shows how non-job variables can and do affect adaptation to the job. This is a list of stresses with assigned numerical values—the highest of 100 being the death of a spouse—also includes some occupational items such as job loss. A high score on that scale covering experiences during the previous year correlates with accident frequency and severity, with the development of disability and with decreased work performance.

The problem is even more complex when one considers the importance of non-occupational variables in reviewing the research on occupational stress. For instance many studies compare the characteristics of different specialties within a broad field. Medicine is a good example. The comparison of surgeons with psychiatrists or with gynecologists makes the assumption that, since they are all physicians, all the job variables are essentially the same. Clearly one cannot do so. The very factors which lead a physician to choose a specialty may be those which are subsequently measured rather than those which relate to how a gynecologist, a psychiatrist or a surgeon reacts to specific stressful events.

Further support for this viewpoint comes from a study by McLean and Hupalowsky. They selected a group of patients on the basis of the presence of a psychiatric illness sufficiently severe to cause work absence. A second criteria for study was the report of a change in employment status just prior to the onset of symptoms. Job factors which triggered the symptoms ranged from retirement without an expected preretirement promotion to promotion itself. In 25 percent the triggering factor was a demotion.

But the apparently job-stimulated disabilities in their cases were invariably a final stressor at the time an individual was adjusting to an unrelated personal problem or physical illness. In this group, when external influences had strained the individual's ability to adapt, he apparently turned for emotional support to his close relationship with his company; when that relationship became strained, he developed psychiatric symptoms. Non-occupational variables then must be considered both in clinical situations and in research.

PSYCHODYNAMICS

In considering that which may be stressful to an individual at work, the worker's mental *set* has been shown to be of major importance. Dr. West cited U.S. Air Force experiments involving the use of subjects in a human centrifuge. Very simply, the psychophysiological stress reaction was measured by the subjects, blacking out as the "g" forces increased. The point at which subjects succumbed to these forces could be easily influenced by altering their emotional state. If they were made by the researchers to feel extremely anxious and uncomfortable, the blackout came much earlier than with the group of control subjects. If on the other hand they were made to feel extremely angry, it took much more "g" force to cause a blackout. Nothing in the research, however, made clear what the psychophysiological mediating forces might be.

The whole emotional tone an individual experiences at any point in time similarly determines to a large degree both the factors to which he will react and the degree and kind of reaction. Less than clearly conscious determinants are involved. If, for instance, an individual is told under hypnosis that the burn caused by a specific, known (temperature, size) hot object will not be particularly painful, there will be much less of a blister, much less erythema, than if it were suggested to him that his skin was already sore or that he would get a severe reaction from the same stimulus. (The difference between guilt and shame as a response to a situation further illustrates this point. If one responds to an embarrassing situation with shame, the physiological reaction is one of flushing; with guilt skin turns pale and one tends to sweat.)

The mental set, occasioned by one's total present, past and recent real life situation on the job and off is vital to the consideration of any occupational stress reaction. Related to this concept is the emotional or intellectual disability (mental illness or psychopathology) which may be influencing a worker's behavior. Many studies have demonstrated the severity and the incidence of psychiatric illness in a normal population. It is much greater than one would ordinarily assume. These varying degrees of disability are brought to the job and obviously affect work adjustment. Today a very common example, *alcoholism*, correlates highly with decrements in performance on the job, accidents, homicides and suicides. With the vast number of problem drinkers in our employed population we may look upon this condition as a stress the individual brings to bear on the job situation itself. Other patterns of overt disability cover the entire range of psychiatric disorder.

Maturation

It is important to recognize that individuals have different maturation rates on jobs and the jobs themselves call for differing degrees of experience before they are effectively performed with reasonable levels of stressfulness. Implicit in the concept is the whole idea of the adaptation of an individual to his job which in turn begs the question of adequate selection and job assignment in keeping with personality characteristics, ambitions and desires. One person likes independence and functions well with it; a second needs structure. Such relationships should receive more consideration than they have in most job placement settings.

Related to this is the concept of self-selection on the job. To what degree can one correlate factors leading a person to select a specific career with the specific characteristics of the job itself? And when there is a sound match, does this insure fewer reactions to factors which might otherwise be regarded as stressful for a specific individual?

The Value of Stress

Rats exposed to repeated and severe stress are subsequently much *less* apt—when exposed to carcinogens—to develop cancer of the skin than those not exposed to stress. And there are innumerable examples from human experimentation and clinical experience which further suggest that there are uses or values of stress. Those who train for athletic events specify how they get "up" for their particular contest. In short, he "stresses' 'himself—he gets the adrenaline up. And it is interesting to note that a great many Olympic champions became champions only after extensive trauma—often affecting a member used in the athletic contest in which they excel (a foot, an arm, etc.).

One could make a case which suggests that the absence of stress is indeed a stress in itself; a lack of variation in one's life pattern can be a stressor. One may refer to Karl Menninger's *The Vital Balance*, in which he quite logically establishes that too much or too little of *anything* for a particular person at a specific time can be a stressor and can produce, in the province of the engineers, strain.

Real stress on the job—dangerous issues in the work situation—potentially life threatening situations—can in fact provide relief for non-job stress which is not environmental but which stems from inner conflict. The sense of relief in a stressful work role tends to relieve an individual of the necessity of reflecting on what's really bothering him.

Many people appear to constantly seek out life threatening situations which would be regarded as stressful. The miner, for instance, in constant danger, occasionally flaunts the rule against smoking underground. And in policemen, the leading cause of death is automobile accidents on and off duty. Why?

Dr. West speculated that there may be two reasons. The first is the need to negate job-related risks by self-induced risk. There are dangers on the job over which one has no control. One way to cope with these is to expose oneself to dangers over which one does have some control. That, in some way, may make it easier for an individual to face the job-related risks. Secondly, and a part of what we may think of as a counter phobic defense, is a desensitization process. By giving oneself a lot of stress—by working one's adrenals overtime—then when one comes to grips with stress on the job it is much eaesier to handle.

To Cope

Finally, considering the natural things that people do to escape from stressfulness in their lives, Dr. West concluded, and other members of the conference were generally quick to agree, that the concepts briefly enumerated below were, singly and together, possessed of considerable value in assisting individuals to cope with factors they found stressful.

Casting out stress. This concept relates to the currently popular acceptance of transcendental and other forms of meditation. Such meditation should not be rejected out of hand by mental health professionals.

Substitution of vicarious stress for real life situations. The tremendous amount of spectator activity in the world today, the increased amount of dependence on drama, on television, and on other people's coping with stress may be therapeutic behavior.

Physical activity. Exercise and sports to stimulate greater physical health appear to be associated with better ability to handle stress.

The recreation explosion. The way more and more people are seeking out alternate activities beyond work seems related to insuring more successful adaptation to stress.

The relationship of natural biological rhythm to the environmental circumstances of a job. Whether it be the menstrual cycle or diurnal variation, different people have different cycles and these should be carefully evaluated.

Finally, and related to the last, ask the worker his proclivities, his desires, his interests. Ask him what kind of a person he is. Is he a morning person, an afternoon person, or an evening person. What does he like to do? What has he done well? Honor this information in job placement if at all possible.

CHAPTER 4

A PSYCHOANALYTIC FRAMEWORK*

OF THE MANY clinical concepts of stress discussed at this conference, those which had their roots in unconscious determinism were widely accepted. There was general agreement that a major source of stress reaction lies in feelings which are difficult to fully assess or understand. The clinicians argued that often an individual does not really know *what* he is feeling, and certainly much of the time is unaware of *why* he feels as he does.

At the conference, Dr. Levinson offered as an integrative frame of reference a theoretical model based largely on psychoanalytic concepts. He talked of a compelling need for such integrated thinking. He identified the need for overarching rubrics, bridging from one set of variables and concepts to another. A simple association or a slight correlation of certain physiological or behavioral variables with certain work-related behavior does not, he felt, support comfortable conclusions about their relationship. The need for spelling out the intervening processes is clear, and Dr. Levinson demonstrates that these processes have not heretofore been clearly established. He says there is no way to explain how these forces act to produce the correlations, and without this understanding we are clearly in the dark.

At this point Levinson offered an interpretation of Freudian theory as such a framework for integration. He pointed out that fundamental psychoanalysis approaches the understanding of work behavior from five points of view.

---

*The informal presentation of Dr. Harry Levinson was extremely detailed and comprehensive. It appears under the title "A Psychoanalytic View of Occupational Stress" in Volume 3, Number 2 (Summer) issue of *Occupational Mental Health*. From his notes of the original presentation, the editor has paraphrased selected highlights to document their inclusion in the conference.

1. The drive theory or instinct theory.
2. A structural conception including id, ego, and superego.
3. Levels of consciousness.
4. A developmental conception of personality (stages of development).
5. Concepts of adaptation among all these forces (to each other and to the environment).

"When we talk about psychological stress on the job we are talking about the disruption of this balance which produces certain kinds of reactions. Sources of disruption (psychologically speaking) have to do with fear, guilt and anxiety tolerance," he said.

One source of dysfunction occurs when the drives are frustrated and not allowed to be discharged constructively. One can therefore look at stress as something which the ego has great difficulty in handling. He suggested we also see a crumbling superego in the face of stress producing difficulty in coping. A variety of occupational stresses can produce a lowering of self-esteem, such as when one's functioning on the job falls far short of one's ego ideal. The same thing happens when people age and no longer have the skills and competence or the flexibility to cope successfully and when there is substantial loss such as when colleagues with whom one has worked closely are transferred or otherwise leave. Loss of support for the ego and superego, such as is involved in a loss of skills or a loss of capacity to adapt, can be another cause for stress reactions, Dr. Levinson suggested.

Loss of support from the work organization with which one has identified and become deeply ego involved comes about through loss of position, retirement, transfer, or organizational failure. This is another clearcut stress to the ego and its supports. People go to work for various organizations for many psychological reasons; some are attracted to AT&T, some to IBM. This provides people with a supporting context and an accepted set of values which dovetail with individual personal values and needs.

Superego conflict may occur when people are asked to behave in ways which violate their own internal standards as, for example, in producing shoddy products, misrepresenting goods and services, or carrying out company policies and practices inimical to others.

The guilt which results is readily evident. Dr. Levinson reported seeing examples of guilt induced behavior when an individual had leaked government information.

Dr. Levinson suggested there were few fruitful avenues of research on occupational mental health utilizing concepts relating to levels of consciousness. On developmental issues, however, he feels that one can begin to look at some of the difficulties which arise from that stress which is either a consequence of fixation at a given level of development or produces a regression to a given level of development. He suggested as an example, that one occasionally sees young men in a work situation who are asked to merely wait in line for their managerial places and promised the opportunity to act in their later years—a time more conducive to reflection than to action. And he suggested this was a very serious bind. In addition, certain occupations are specifically related to life stages and are marked by transitional difficulties. Professional athletes, for instance, notoriously must deal with this problem. So must scientists, creative artists and many others who find themselves in occupational stress situations related to specific life stages.

Finally, threats to adaptation are consistently stressful. When the unconscious psychological contract between a person and his work organization is threatened, a stress reaction is the common result. People who have worked hard, who have done what is expected of them in their company and who suddenly find themselves out of a job, feel they have been done an injustice no matter what the economic reasons for the action.

Dr. Levinson said that the adaptive task for people in management ranks involves increasing intensity of demand on their employees for performance, particularly in organizations whose heads are pre-occupied with running them for quarterly reports to security analysts. Such a focus, he said, inevitably means that people are moved up and down an organizational hierarchy like yo-yos with increasing pressure to make the results look good. Organizations, like competitive children, must look good to those whose approval they require.

That pressure in turn leads to putting people in occupational roles which they experience as overwhelming because they simply do not have the capacity to adapt to them. Sometimes this is done because a person has been remarkably successful in one role. On the basis of that success he has moved to another role wherein he fails.

Adaptation involves coping with one's own character defenses in the face of tasks that are age-specific, stage-specific, and congenial to one's equilibrium maintaining efforts. Adaptation involves managing the drives, managing the distance between the ego ideal and the self-image, and maintaining one's focal conflicts in the face of institutional requirements which may magnify or intensify them, or conversely make them incongruous in given situations. These issues must be taken into account by any psychologically sophisticated management. These are dimensions of occupational stress which require our careful attention.

~~~~~~

CHAPTER 5

STRESS, DISTRESS AND PSYCHOSOCIAL STIMULI

LENNART LEVI

THERE IS NO QUESTION but what environmental *physical stimuli* cause disease. This has been well established for a large number or factors and diseases. As the exposure, avoidance or manipulation of physical factors increases or decreases, the chance of becoming ill or reversing ill health occurs.

The role of *psychosocial stimuli* is not so clear. This chapter will present a perspective on the relationship between such stimuli and various factors associated with physical disease. First, it will focus on theoretical concepts which relate to the development of psycho-somatic disorders. Next, it will illustrate various aspects of these matters with data from recent studies. Finally, it will present some viewpoints on the prevention of psychosomatic disorder.

PSYCHOSOCIAL STIMULI, STRESS AND DISEASE

Stress will be a major consideration in this discussion. The term "stress" is used as Selye has used it. That is, the *non-specific response of the body to any demand made upon it* (Selye, 1971). The intensity and duration of this stereotyped, phylogenetically old adaptation pattern which prepares the organism for fight or flight is assumed to be closely related to the rate of wear and tear in the organism. Consequently it is probably also related to the morbidity and mortality. Stress as used here is not only related to any *specific* disease but is probably associated with a variety of diseases. In other words, if environmental changes occur frequently and/or are of great magnitude and/or if the organism is particularly vulnerable, stress reactions usually increase in intensity and duration. This is shown in Figure

[31]

5-1 which demonstrates the hypothetical relationship between various levels of environmental stimulation and the resultant amount of "stress Selye."

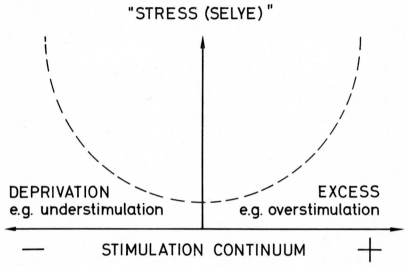

Figure 5-1. Theoretical model of the relation between physiological stress (as defined by Selye) and various levels of stimulation. According to this hypothesis, deprivation of stimuli as well as excess is accompanied by an increase in "stress (Selye)", (Levi, 1972).

Hypochondriasis, Functional Disturbance, and Structural Damage

Clinically, overt symptoms may appear even with relatively low stress levels. This occurs, for instance, if a patient pays untoward attention to sensations originating from such reactions. Thus, in some individuals and under certain circumstances, even "normal" sensations of this type may be interpreted as symptoms or disease (as in the case of hypochondriasis).

If psychosocial stimulation is pronounced, prolonged or often repeated, and/or if the organism is predisposed to react because of the presence or absence of certain interacting variables, the result may be a hyper-, hypo-, or dysfunction in one or more organs, not only in those subjects who are particularly aware of and sensitive to their proprioceptive signals but in perfectly "normal" subjects as well. Examples of such reactions are tachycardia, syncope, pain of vasomotor or muscular origin, hyperventilation, and increased or decreased gastrointestinal peristalsis.

Almost by definition, these symptoms are detrimental to the normal *function* of the organism, as in the case of an organ neurosis. If such reactions are particularly prolonged, there is suggestive evidence that they may eventually lead to *structural* damage as well.

Given these considerations let us now examine some of the principal hypotheses in the study of psychosocially mediated disease. (See Fig. 5-2). The main hypothesis is that psychosocial stimuli can cause disease. The etiology may be specific—that is a *specific* disease may be caused by a particular stimulus. The cause can also be *non-specific*. That is, disease will occur in response to a wide variety of stimuli and/or in a wide variety of subjects.

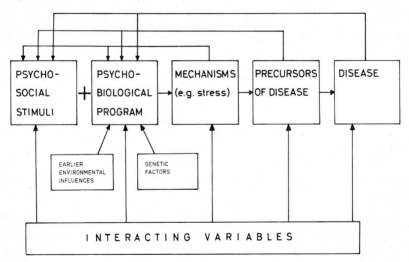

Figure 5-2. A theoretical model of psychosocially mediated disease. The combined effect of psychosocial stimuli (1) and man's psychobiological program (2) determines the psychological and physiological reactions (mechanisms) (3), e.g. "stress (Selye)" of each individual. These may, under certain circumstances, lead to precursors of disease (4) and to disease itself (5). This sequence of events can be promoted or counteracted by interacting variables (6). The sequence is not a one-way process but constitutes part of a cybernetic system with continuous feed-back (Kagan and Levi, 1971).

Hypotheses on Psychosocially Mediated Disease

In the present context I will focus on the *non-specific* etiology and will suggest hypotheses concerning disease that are psychosocially mediated:

1. Most life changes evoke a stereotyped, phylogenetically old, adaptational response pattern which prepares the organism for physical activity, e.g. for fight or flight.
2. "Stress (Selye)" is characterized by increased activity in the hypothalamo-sympathoadrenomedullary, hypophyseo-adrenocortical, and possibly the thyroidal system.
3. The reactions, at least if prolonged, intense or often repeated, are accompanied by an increased rate of wear and tear in the organism.
4. Such generally increased rate of wear and tear leads, in the long run, to increased morbidity and mortality.

Figure 5-2 demonstrates a schematic overview that suggests external influences (identified as "psychosocial stimuli") combined with what may be thought of as constitutional factors, e.g. personality variables (labeled as the "psychobiological program"). They are together productive of reactions, mechanisms, e.g. like "stress (Selye)." These in turn may provoke precursors of disease and, continuing in evidence, disease itself.

As in the case of all conceptual models, this one, too, implies a gross oversimplification of highly complex issues.

Review of Evidence

To be concise, the supporting evidence will be summarized in the following figures. Psychosocial stimuli lasting for hours or days or months have all been associated with biochemical reactions associated with "stress (Selye)."

BIOCHEMICAL REACTIONS IN RESPONSE TO PSYCHOSOCIAL STIMULI. Figures 5-3 and 5-4 A and B demonstrate that such stimuli lasting for some hours increase adrenaline and nonadrenaline excretion and also the levels of plasma free fatty acids and triglycerides.

Psychosocial stimuli lasting several days also increase "stress (Selye)." This is reflected in the increase in adrenaline excretion noted in Figure 5-5 and in the plasma free fatty acids and cholesterol

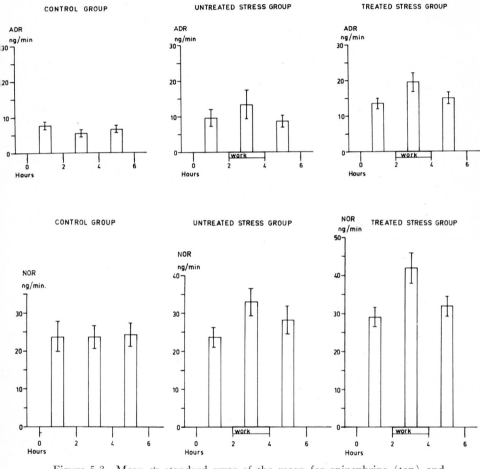

Figure 5-3. Mean ± standard error of the mean for epinephrine (top) and norepinephrine (bottom) excretion, under conditions designed to induce distress under the second of the three 2-hour periods in the untreated (center) and nicotinic acid treated (right) stressor-exposed groups but not in the control group (left), (Carlson, *et al.*, 1972).

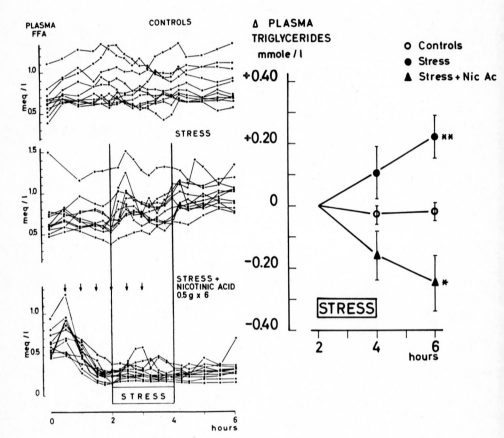

Figure 5-4. A: Individual values for arterial plasma levels of free fatty acids in the control group (top), the untreated stressor-exposed group (center), and the treated stressor-exposed group (bottom). Arrows indicate nicotinic acid administration (0.5 g 6 times, i.e. every thirty minutes in the treated stressor-exposed group (Carlson, *et al.,* 1972).

B: Mean ± standard error of the mean for the changes in plasma tri-glycerides during and after the second 2-hour period, which was designed to induce distress in the untreated and treated stressor-exposed group but not in the control group. * and ** indicate that p < 0.05 and 0.01, respectively (Carlson, *et al.,* 1972).

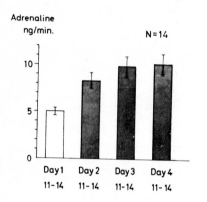

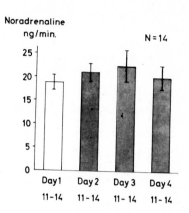

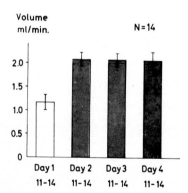

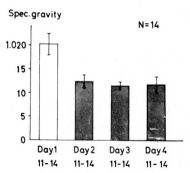

Figure 5-5. Urinary adrenaline and noradrenaline excretion, urine flow, and specific gravity during control conditions (empty bars) and corresponding periods of days 2, 3 and 4, at eleven to fourteen hours. Means ± S.E.M. (Levi, 1972).

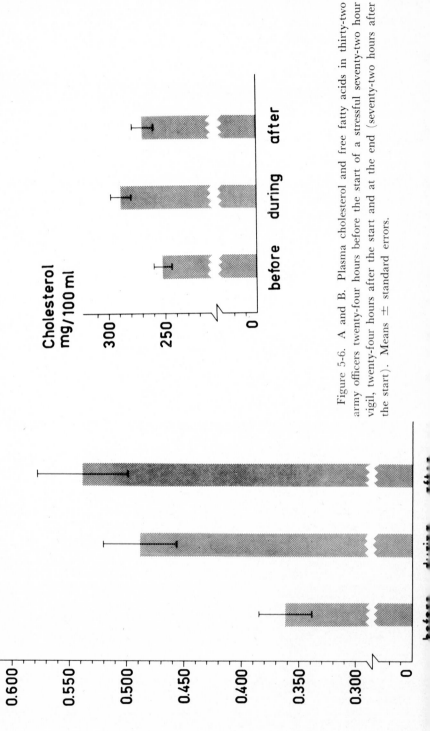

Figure 5-6. A and B. Plasma cholesterol and free fatty acids in thirty-two army officers twenty-four hours before the start of a stressful seventy-two hour vigil, twenty-four hours after the start and at the end (seventy-two hours after the start). Means ± standard errors.

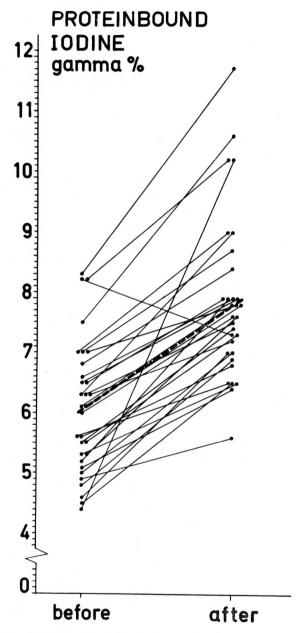

Figure 5-7. Protein-bound iodine before and after a seventy-two-hour vigil (Levi, 1972).

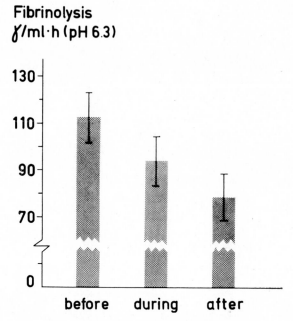

Figure 5.8. Fibrinolysis in thirty-two army officers twenty-four hours before the start of a stressful seventy-two-hour vigil, twenty-four hours after the start and at the end. Means ± standard errors.

which are shown to be elevated in Figure 5-6 A and B. The protein-bound iodine in Figure 5-7 and the fibrinolysis levels demonstrated in Figure 5-8 support this hypothesis.

Longer-acting psychosocial stimuli also increase "stress (Selye)." The evidence in support of this hypothesis is drawn from a longitudinal study demonstrating a significant, positive correlation between the reported weekly sum of "life change units" and adrenaline excretion during the days prior to interviews (Rahe, 1969).

The studies all support the hypothesized relationship between psychosocial stimuli (Figure 5-2) and potentially pathogenic mechanisms.

Psychosocial Stimuli as Related to Disease. Various life changes may be regarded as psychosocial stimuli. And these, if frequent and/or intense, have been shown to precede an increase in morbidity and mortality in a *variety* of illnesses, psychiatric as well as somatic. Holmes, Rahe and others have demonstrated that the greater the degree of change in a subject's life during a certain period of time, the higher is his risk of developing subsequent decrease in health status (Holmes and Rahe, 1967). It may be noted that the life changes under consideration also include those usually considered pleasant such as an engagement, a marriage, and the gaining of a new family member.

A number of studies supporting this relationship between individual cumulative life changes with subsequent nonspecific health changes have been published. Among the more recent are those of Theorell (1970), and Rahe (1972) and Theorell, Lind *et al.* (1971).

There is evidence of a relationship between psychosocial stimuli and some specific diseases including thyrotoxicosis, some cardiac disorders, essential hypertension, and peptic ulcer. (For a review, see Kagan and Levi, 1971.) This evidence tends to support the relationship hypothesized in Figure 5-2.

RELATIONSHIP BETWEEN MECHANISMS AND DISEASE. Increased levels of triglycerides and cholesterol predict increased risk not only for subsequent ischemic heart disease (Keys, 1970) but also for a variety of other diseases as found by Tibblin in a large scale prospective study on males born in 1913 and followed annually since 1963. (See Figs. 5-9 and 5-10.) These findings support the relationship hypothesized in Figure 5-2 between the boxes labeled "mechanisms" and "precursors of disease" and that called "disease."

Furthermore, various diseases have been induced experimentally in animals, including primates, by exposing them to psychosocial stressors. In this way, several authors induced degenerative heart disease and hypertension (Lapin and Cherkovich, 1971). Further supporting the hypothesis of psychological stimuli as possible etiological agents in a variety of diseases.

POTENTIALLY NOXIOUS PSYCHOSOCIAL ENVIRONMENTAL CONDITIONS. From this we may conclude that environmental conditions may provoke stereotyped reactions, involving changes in bodily functioning which may be rather pronounced. These reactions include increases in catecholamine excretion and in plasma levels of free fatty acids, triglycerides, cholesterol and protein-bound iodine. They are also associated with a decrease in serum iron and fibrinolysis. The increased catecholamine output, occurring in response to almost any type of environmental change is correlated positively with the intensity of self-reported distress.

It has further been shown that subjects exposed to frequent and

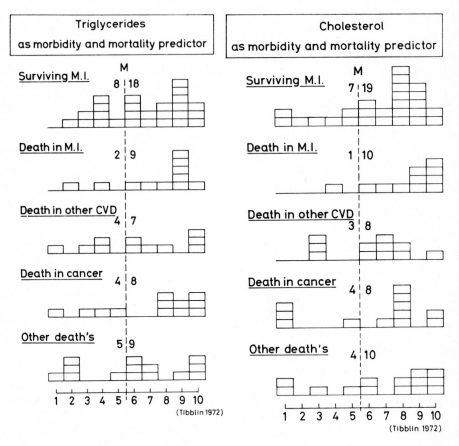

Figures 5-9 and 5-10. Plasma triglycerides and cholesterol as predictors of morbidity and mortality in myocardial infarction (M.I.), and of mortality in other diseases. Distribution in deciles (1-10) of plasma levels in relation to deaths. Vertical line indicates median. Each box indicates one death or case of illness. Figures indicate distribution of below-median (left) and above-median (right) triglyceride and cholesterol levels (Tibblin, personal communication). Above-median levels seem to predict subsequent morbidity and mortality (Levi, 1972).

dramatic environmental changes run a higher risk of developing myocardial infarction within the next six months (Theorell, 1970). Similar relationships have been demonstrated for a variety of somatic and psychiatric diseases (Rahe, 1972).

Finally, it has been found that high triglyceride and cholesterol levels in a random population are positively related to a subsequent increase in mortality in a variety of disorders.

Taken together, the studies demonstrate the probable importance of non-specific psychosocial environmental stimuli and non-specific physiological mechanisms in the production of a variety of diseases. We further conclude that the causation of disease by such stimuli is not proven, but that there is a rather high level of suspicion. Nor are we saying that specific factors are unimportant. Both deserve and need further study, focusing on a number of psychosocial environmental factors.

MONITORING AND PREVENTION
PRACTICAL IMPLICATIONS

Before one can discuss prevention in scientific terms it is necessary to have more information. This means research, and research means both planning and measuring.

But health administrators will continue to feel that action should be taken based on the existing levels of suspicion. Whether such action is in the area of marriage counseling, crisis intervention, developing more satisfying jobs, or of programs for the aging, it is mandatory to regard the actions as only a trial. Concurrently, means must be established of evaluating the program for efficiency, safety and cost. This has both a scientific and a social purpose; the former to establish knowledge; the latter to protect individuals from danger. It also serves to avoid a sense of false security, to prevent delay in applying useful procedures, and to provide rational support for innovative measures based on known cause and effect.

Since our knowledge of psychosocial hazards and disease is relatively incomplete, much of what we say about *monitoring* for and *preventing* disease is speculative. Changes likely to expose large numbers of people to new social relationships or generally requiring them to make major adaptations may result in subsequent health hazards (Kagan and Levi, 1971). Awareness of this possibility should be in the minds of the leadership of any organization contemplating such change whether in a work organization or at the community level. If change is likely to be accompanied by removal from old forms of social support and either lack of, or failure to use, new forms, there might be a need for preventive action.

Awareness of possible hazards in any group—large or small—may arise from the examination of morbidity records. Such data are not likely to be conclusive in any way, but high and/or rapidly increasing rates for suicide, neuroses, and psychosomatic disorders may provide a community with suggestions for more detailed inquiry. In the same way increases in sickness and accident benefits, the use of health insurance, and accident rates may suggest to the leadership of a work organization the need for investigation and the establishment of techniques of prevention. Such prevention could focus on rather general phenomena in large groups of subjects; that is, the principal focus could be on deprivation or excess of a large number of stimuli. Concerns could be with such matters as over- or understimulation, lack of parental care or overprotection, poverty or affluence, isolation or a lack of privacy, and restriction of action or extreme permissiveness. (Thus *prevention* could be focused on rather *general* phenomena in large groups of subjects.)

A detailed discussion of primary and secondary prevention is not possible here. The identification of some general principles is. Based on the theoretical model presented (Figure 5-2) and on the non-specific hypotheses discussed, one may note five principles, or perhaps more accurately, practices, of prevention:

1. Intervening against psychosocial stressors.
2. Minimizing predisposing and stimulating protective interacting variables.
3. Intervention in the area of "mechanisms."
4. Intervention against precursors of disease.
5. Secondary prevention.

While some preventive measures may be non-specifically beneficial to all individuals and under all or most conditions, in many instances it is necessary to stipulate the specific context, the particular target group, and the specific disease entity with which one is concerned.

Much preventive action which is assumed to be beneficial will be promoted by politicians and health and welfare administrators without waiting for evidence from scientists. Under such conditions this chapter advocates evaluative research, turning every major social policy act into a large scale experiment. By making our social system cybernetic, with feedback loops, our social policies may eventually become self-corrective so that major hazards originating from exposure of large populations to environmental stressors and/or to well-meant, but badly founded, social action can be eliminated or at least modified.

It must be further emphasized that adverse effects upon an individual's physical or mental health do not *ipso facto* suggest a need to alter the psychosocial or other stimuli producting this reaction. It may well be that some ill effects are outweighed by advantages (economic, social or psychological) for the individual or the collective community. The balancing of debts and credits is not and can never be a purely medical affair. It must be subjected to political assessment. Briefly, then, we must define our criteria as to what to prevent and why. We must clarify what price we are willing to pay for disease prevention. On the other hand, decision-makers must not forget the potential price in their actions which may be paid in terms of health and well-being.

Finally, in many of today's highly urbanized and industrialized societies, we tend to forget that efficiency is a means, not an end. If we are asked to choose between producing more and better goods at the expense of our social integrity and producing fewer goods of lower quality, we will unhesitatingly choose the latter if by such choice we can avoid pain and disaster to our people.

BIBLIOGRAPHY

Brown, G. W., Harris, T. O., and Peto, J.: Life events and psychiatric disorders. II. Nature of causal link. *Psychol Med,* 1972, in press.

Brown, G. W., Sklair, F., Harris, T.,O., and Birley, J.L.T.: Life events and psychiatric disorders. I. Some methodological issues. *Psychol Med,* 1972, in press.

Carlson, L. A., Levi, L., and Oro, L.: Stressor-induced changes in plasma lipids and urinary excertion of catecholamines, and their modification by nicotinic acid. In Levi, L. (Ed.): *Stress and Distress in Response to Psychosocial Stimuli.* Oxford, Pergamon, 1972, pp. 91-105.

Holmes, T. H., and Rahe, R. H.: Social adjustment rating scale. *J Psychosom Res, 11:*213, 1967.

Kagan, A. R., and Levi, L.: Health and environment—psychosocial stimuli. A review. *Rep Lab Clin Stress Res (Stockholm,)* 27, 1971.

Keys, A.: *Coronary Heart Disease in Seven Countries.* New York, American Heart Association, 1970.

Lapin, B. A. and Cherkovich, G. M.: Environmental change causing the development of neuroses and corticovisceral pathology in monkeys. In Levi, L. (Ed.): *Society, Stress and Disease—The Psychosocial Environment and Psychosomatic Diseases.* London, Oxford University Press, 1971, vol. 1, pp. 266-279.

Levi, L.: *Stress and Distress in Response to Psychosocial Stimuli.* Oxford, Pergamon, 1972.

Paykel, E. S., Myers, J. K., Dienfelt, M. N., Klerman, G. L., Lindenthal, J. J., and Pepper, M. P.: Life events and depression: A controlled study. *Arch Gen Psychiatry, 21*:753, 1972.

Rahe, R. H.: Life crisis and health change. In May, P.R.A., and Wittenborn, J. R. (Eds.): *Psychotropic Drug Response: Advance in Prediction.* Springfield, Thomas, 1969.

Rahe, R. H.: Subjects' recent life changes and their near-future illness susceptibility. *Adv Psychosom Med,* 8, 1972.

Rioch, D. McK.: The development of gastrointestinal lesions in monkeys. In Levi, E. (Ed.): *Society, Stress and Disease—The Psychosocial Environment and Psychosomatic Disease.* London, Oxford University Press, 1971, vol. I, pp. 261-265.

Selye, H.: The evaluation of the stress concept—stress and cardiovascular disease. In Levi, L. (Ed.): *Society, Stress and Disease—The Psychosocial Environment and Psychosomatic Diseases.* London, Oxford University Press, 1971, vol. I, pp, 299-311.

Theorell, T.: *Psychosocial Factors in Relation to the Onset of Myocardial Infarction and to Some Metabolic Variables—A Pilot Study.* Thesis, Department of Medicine, Seraphimer Hospital, Stockholm, 1970.

Theorell, T., Lind, E., Froberg, J., Karlson, C. G., and Levi, L.: A longitudinal study of twenty-one subjects with coronary heart disease—life changes, catecholamine excretion, and related biochemical reaction. *Psychosom Med, 34*:505-516, 1972. Summary in *Psychosom Med, 33*:465, 1971.

CHAPTER 6

CONFLICT, AMBIGUITY, AND OVERLOAD: THREE ELEMENTS IN JOB STRESS

Robert L. Kahn

L ARGE SCALE ORGANIZATIONS have become a subject of sustained interest to social scientists, and no wonder. Much of life is lived in organizations, and the quality of extra-organizational life is to a considerable extent organizationally determined. The Gross National Product is essentially an organizational product.

These facts correspond to two rather different lines of organizational research, theory and application. One has to do with productivity. Research in this tradition looks for differences between effective and ineffective organizations, high-producing and low-producing work groups, successful and unsuccessful supervisors, "motivated" and "unmotivated" workers. Theories that direct or accrue from such research concentrate on the explanation of organizational effectiveness. Application centers on supervisory training, worker exhortation, and organizational development.

More recently, research has begun on the other major aspect of organizations—their human membership and the effects of organizational demands and opportunities on individual members. Research in this still embryonic tradition seeks to explain occupational and other role-related differences in health or illness, mental and physical. Similar research aims have also generated studies that identify some hypothesized form of stress in organizations and attempt to trace its effects on individuals.

Theory in this context is likely to take its criterion variables (effects) from medicine, psychiatry, or clinical psychology, and its independent variables (causes) from sociology or organizational psychology. Application is at too primitive a stage to encourage generalization, but

[47]

some practitioners of organizational development are attempting to synthesize the attainment of organizational effectiveness and individual well-being. The view of organizations as socio-technical systems (Trist and Bamforth 1951; Rice 1958; Trist, *et al.,* 1963) and the recent emphasis on the humanization of work exemplify the growing concern with the effects of organizations on individuals.

That concern is at the core of our research program—Social Environment and Mental Health—now in its fourteenth year at the Institute for Social Research of the University of Michigan. The broad aims of the program are expressed in the explanatory framework shown as Figure 6-1. The figure specifies immediately the six sets of variables around which the program is constructed; they appear as the numbered boxes in the figure. The categories of relationships and hypotheses are represented by arrows, which also remind us of the directions of causality to be emphasized.

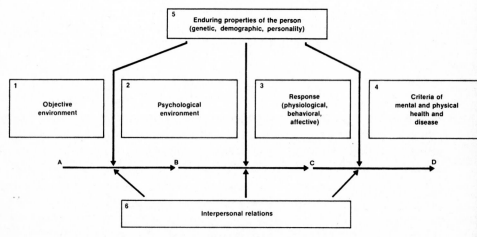

Figure 6-1.

Thus, hypotheses of the A→ B category have to do with the effects of the objective environment on the psychological environment (the environment as the individual experiences it). To take an example within the subject of job stress, we find that people whose jobs require them to engage in transactions across the organizational boundary (a fact in their objective environment) more often report that they are subjected to incompatible demands (a fact in their psychological environment). Hypotheses of the B→ C category relate facts in the psychological environment to the immediate responses that are invoked in the person. For example, the perception that one is subject to persistent conflicting demands on the job is associated with feelings of tension. The C→ D category deals with the effect of such responses on criteria of health and illness. The relationship of job tension to coronary heart disease illustrates the C→ D category.

Finally, the categories of hypotheses just described must be qualified by an additional class, represented by the vertical arrows in the figure. This class of hypotheses states that relationships between objective and psychological environment, between psychological environment and response, and between response and criteria of health are modified by enduring characteristics of the individual and by interpersonal relations. For example, the extent to which a person experiences tension on being exposed to role conflict depends very much upon the personality characteristic of flexibility-rigidity; people who are flexible rather than rigid respond with greater tension to the experience of role conflict. In similar fashion, other properties of the person and his interpersonal relations act as conditioning variables in the hypotheses described above.

The research aims of our program are represented by all four of these categories of hypotheses, in combination. We are attempting to develop a theory of mental health as it is affected by the contemporary environment of the individual, taking into account the facts of genetic endowment and personality insofar as necessary to make sense of the environmental effects. Given this framework, it is clear what constitutes an adequate explanatory sequence—a chain of related hypotheses beginning with some characteristic of the objective environment, ending with some criterion of health, specifying the intervening variables in the psychological environment and response categories, and stating the ways in which this causal linkage is modified by the differing characteristics of individuals and their interpersonal relations. Such a causal chain we call a *theme*. We have attempted to state the aims of each research project in terms of one or more such themes, and to make the results of research projects additive by building toward the completion of such themes.

In conducting such thematic or programmatic research, we find that successive projects build on their predecessors in five main ways—by defining more precisely or differentiating concepts, by improving measures, by demonstrating more clearly (or extending further) some effect under study, by specifying more completely the relevant conditions or populations for which a given cause-and-effect relationship holds, and by extending the demonstrated relationships earlier into the causal sequence (discovering factors antecedent to those already known). This paper and those by Dr. Cobb and Dr. French illustrate all these processes for a central concept in our program—job stress. We will be particularly concerned, however, to show the differentiation of the stress concept through a sequence of a dozen or more research projects.

I shall be referring to six of these projects, and will describe them now; hereafter I will mention them only by name or convenient abbreviation:

1. *Conflict and ambiguity, intensive study:* This study—by Kahn, Wolfe, Quinn and others (1964)—was built around fifty-three people, selected from several major corporations to represent the full range of jobs from first-level supervisor to corporate officer. Data were collected by interview, written questionnaire, and personality test from each of the fifty-three. Interviews were also conducted with 381 other persons who had been identified as members of the role sets of the fifty-three "focal persons." These 381 people held jobs that made them functionally interdependent with the fifty-three focal persons. In a sense, therefore, the expectations and demands of the 381 defined the roles of the fifty-three. Our purpose was to explore the degree of conflict or harmony, ambiguity or clarity in the role requirements confronted by the fifty-three focal persons and to discover some of the organizational causes and individual consequences.

2. *Conflict and ambiguity, national survey:* This study (Kahn, et al., 1964), done in conjunction with the study or role sets just described, was an interview survey of about 1,500 respondents. Each of them was treated as a focal person and responded to the same basic questions about his work role as did the fifty-three focal persons in the intensive study. The main contribution of the national survey, of course, was information about the prevalence of role conflict and ambiguity in the work situation and replication of some of the findings of the intensive study on a representative population. The obvious

methodological limitation of the national survey was the reporting of both causes and effects by individual respondents; no data were obtainable from members of their role sets, the people with whom they interacted on the job.

3. *Sales office study* (Kraut, 1965): This study was conducted in the sales department of a large corporation with a decentralized nationwide structure of sales offices. In each of 151 such offices, data were obtained from written questionnaires filled by the sales manager and a sample of salesmen, 823 of whom were included in the research. This study not only replicated major findings from the first intensive study of conflict and ambiguity, but also allowed the collection of quantitative data (in terms of dollars of sales) indicating the magnitude of disagreement and their effects on performance.

4. *Secondary analysis role conflict data* (Sales,, 1969): Several related studies were conducted by Sales, one involving secondary analysis of the earliest study in this series for the purpose of differentiating the concept of role conflict. This analysis concerns us directly. Sales followed this analysis with several experimental studies, with students and with engineers, scientists, and administrators at the National Aeronautics and Space Administration. About eighty-four NASA employes and 150 students were involved.

5. *Goddard Space Flight Center Study* (Caplan, 1971): This study, which included 205 scientists, engineers, and administrators at the Goddard Center (NASA), was designed to press further with overload as a particular form of role conflict and a particular source of stress in certain jobs. The locus of the study and the remarkable cooperation of NASA staff also permitted the collection of individual physiological data not available in any of the previous research.

6. *Kennedy Space Center Study* (French and Vickers): This study was designated primarily as a replication of the work at Goddard Space Flight Center. It also includes more intensive measures of strain and is longitudinal in design. The population consists of about 150 engineers and administrators, for whom there have now been two collections of data. Let us now turn to things learned about role stress from this series of studies:

ROLE STRESS, A GLOBAL CONCEPT

Although we consider the notion of stress important, we use the word as a generic term and have from the beginning sought more specific concepts for our research. In this respect we benefitted from, but did not follow, the leading work of Selye (1956), who defines stress as the reactions of an organism to damaging stimuli. We have adopted instead what Lazarus (1966) has called "the engineering analogy," which regards stress as any force directed at an object, defines strain as the effects of stress, and measures such effects in terms of deflection or some other structural change. To make the metaphor useful, however, we have chosen to study certain specific environmental stresses and equally specific criteria of individual strain.

Role Conflict

The first of these was role conflict, which we thought of initially as consisting of logically incompatible demands made upon an individual by two or more persons whose jobs were functionally interdependent with his own. In the intensive study of role conflict (Study #1) each member of the fifty-three role sets was asked to indicate, for each major job activity of the relevant focal person, the extent to which he wanted the focal person's behavior to change. Four component scores were developed, reflecting responses in terms of amount of change wanted, number of activities involved, changes in time allocation wanted, and changes in behavioral style wanted. These scores, summed for the members of a person's role set, constituted the index of role conflict for him. We regard them as describing a fact in his objective environment, since they represent the views of his role senders rather than his own perceptions; in that sense they are a measure of objective role conflict.

The main effects of such conflict were measured in terms of the responses of the fifty-three focal persons, thus linking panels 1 and 3 in our programmatic schema (Figure 6-1). The effects of role conflict, so measured, are varied in form but consistently negative in their implications for the focal person. Persons subjected to high role conflict report greater job-related tensions, lower job satisfaction, less confidence in the organization itself, and more intense experience of conflict (Table 6-I); these might be considered some of the emotional costs of role conflict.

Similar analyses show that role conflict is also associated with poor interpersonal relations; in comparison to persons subjected to little or no role conflict, persons subjected to high role conflict report that they trust members of their role sets less, respect them less, like them less, and communicate with them less (Tables 6-II and 6-III).

All these represent findings of the A → C pattern in our schema. People who were in high-conflict situations were, naturally enough, more likely to report intense experience of conflict, a finding of the A→ B pattern. However, a strong link between objective and subjective role conflict could not be established because we had no measure of subjective conflict that was commensurate with our measure of objective conflict. The potential link between the focal person's responses and his health was also missing, because in this study we had no adequate measures of health.

We were, however, able to extend in other ways the findings

TABLE 6-I*
EMOTIONAL REACTIONS TO ROLE CONFLICT
(FROM INTENSIVE STUDY)

Emotional Reaction	Degree of Role Conflict		p
	High	Low	
(a) Intensity of experienced conflict	3.3	1.9	<0.07
(b) Job-related tensions	5.1	4.0	<0.03
(c) Job satisfaction	4.4	5.6	<0.02
(d) Confidence in organization	5.7	7.3	<0.001
N	(27)	(26)	

*Reprinted from *Organizational Stress,* Table 4-2.

TABLE 6-II*
INTERPERSONAL CONSEQUENCES OF ROLE CONFLICT
(FROM INTENSIVE STUDY)

Interpersonal Bond	Degree of Role Conflict		p
	High	Low	
Trust in senders	4.5	5.8	<0.01
Respect for senders	4.2	5.9	<0.001
Liking for senders	4.8	5.2	<0.05
N	(27)	(26)	

*Reprinted from *Organizational Stress,* Table 4-3.

TABLE 6-III*
INTERACTIONAL CONSEQUENCES OF ROLE CONFLICT

Interaction Variable	Degree of Role Conflict		p
	High	Low	
Communication frequency	3.9	5.8	<0.001
Power attributed to others	3.8	5.6	<0.001
N	(27)	(26)	

*Reprinted from *Organizational Stress,* Table 4-4.

between objective conflict and individual responses, even in this initial study. One such extension reached further into the objective enviornment, identifying the kinds of positions that were most likely to be characterized by conflicting expectations among role senders. These included positions that required the "crossing" of an organizational boundary—dealing simultaneously with people inside and outside the organization. Positions involving creative problem-solving, in contrast to routine, were also more likely to be conflict-ridden. Positions in supervision and management were more often conflict-laden than were non-supervisory positions.

The responses of individuals to role conflict were not uniform, however; they were mediated or "conditioned" by the personality of the focal person and by the quality of his interpersonal relations. Under high-conflict conditions, people who tended to be anxiety-prone ex-

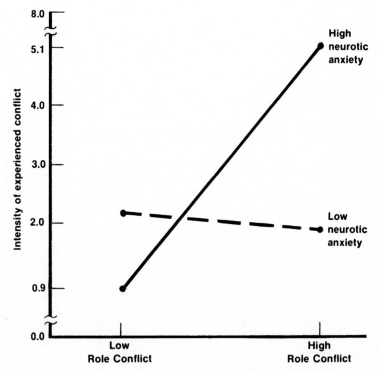

Figure 6-2. Intensity of experienced conflict in relation to role conflict and neurotic anxiety.

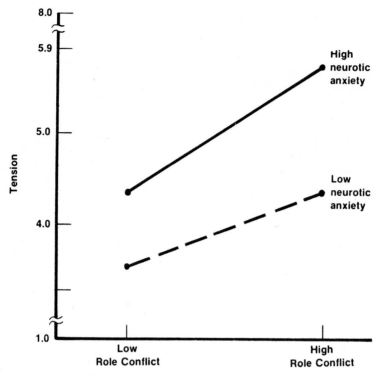

Figure 6-3. Tension in relation to role conflict and neurotic anxiety.

perienced the conflict as more intense and reacted to it with greater tension than people who were not anxiety-prone. (Figures 6-2 and 6-3). Similarly, introverts reacted more negatively to role conflict than did extroverts; they (introverts) suffered more tension and reported more deteriorated interpersonal relations. The personality dimension of flexibility-rigidity mediates still more strongly the relationship between role conflict and tension, with the flexible people accounting for almost the entire effect of role conflict and the rigid people reporting virtually no greater tension in the high-conflict situation than in the low (Figure 6-4).

The relationships between role conflict and various responses indicating strain are mediated by the interpersonal context in much the same way as by the personality characteristics of the focal person. The more frequent the communication between role senders and focal person, the greater the functional dependence of the focal person on

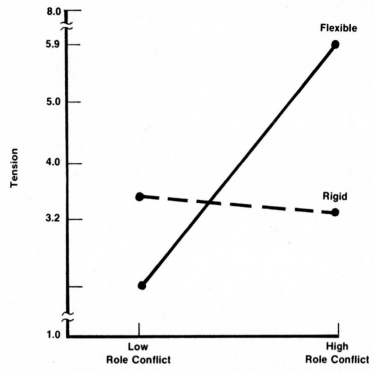

Figure 6-4. Tension in relation to role conflict and flexibility rigidity.

the role set, and the greater the power of the role set over the focal person—the more signs of strain he shows when role conflict occurs. For example, he is more likely to experience a sharp reduction in job satisfaction, a sense of futility, and a negative change in affect for his role set.

The study of fifty-three role sets told us a great deal about the dynamics of role conflict, but very little about its prevalence. The national survey (Study #2) was designed to complement the intensive study by concentrating on the task of description. It revealed that the experience of role conflict is common indeed in the work situation. Almost half of all respondents reported being "caught in the middle" between two conflicting persons or factions, more or less frequently. Most of these conflicts are hierarchical; nine out of ten of the people involved in conflicts reported that at least one of the parties to the conflict was above them in the organization.

Somewhat less than half said that one of the conflicting parties was outside the organization.

These things were not unexpected. We were surprised, however, that one of the more dominant forms of conflict reported did not involve logically or morally incompatible demands, but rather temporally incompatible demands. People complained of overload— willingness to meet the demands and expectations of others, even to acknowledge them as separately legitimate and reasonable, but inability to meet them simultaneously or within the prescribed time limits.

Two of our colleagues, Allen Kraut and Stephen Sales, undertook to extend these research findings, especially by improving and refining the concept of role conflict itself. Kraut's (1965) study, utilizing a population of 823 salesmen and their managers in 151 offices, obtained commensurate measures of objective and subjective role conflict. For example, he measured in dollars the amount of sales that the manager expected from a salesman and the amount that the salesman himself considered appropriate; the difference between the two was taken as an objective measure of conflict. But Kraut also asked each salesman what he thought the manager expected of him; the difference between this amount and the salesman's own estimate of appropriate performance was taken as a measure of subjective conflict. Objective and subjective conflict are significantly correlated, as one would expect; the salesmen were not in a state of delusion. However, they did show a very general tendency to underestimate the manager's expectations, and thus to perceive their supervisors as nearer their own positions than was in fact the case. The effect of this distortion was to understate the conflict; subjective conflict was significantly less than objective conflict. Such distortion can be viewed as a way of coping with conflict, perhaps reducing its tension-evoking effects. It does nothing to resolve the objective problem, and its long-run success in reducing tension must therefore be suspect. Better methods of coping and conflict resolution can be taught.

Role Overload

Stephen Sales (1969) was intrigued by the frequency with which respondents in the first study of role conflict talked about overload when they were asked to discuss the nature of the conflict. He improvised an index of overload from three items from that study—the amount of pressure felt to do more work, the feeling of not being able to finish one's work in an ordinary day, and the feeling that the amount of work interferes with "how well it gets done." He found that these items correlated .6 with job-related tension and explain much of the effects that we had ascribed to role conflict in general.

In subsequent studies at Goddard and Kennedy Space Centers, we concentrated more explicitly on the process of overload and developed separate measures of quantitive load (a continuum running from "having too little to do" to "having too much to do") and qualitative load (a continuum running from "having work that is too easy to do" to "having work that is too difficult to do"). In order to obtain information independent of the person's own perceptions, we supplemented the usual interviews and questionnaires with observations and records of numbers of meetings, office visits, telephone calls, and hours worked.

The distinction between qualitative and quantitative work load had been developed in a study of 122 university professors and administrators (French, Tupper and Mueller 1965). Factor analysis had indicated that these were two distinct and separate variables, and that they showed some similar and some disparate effects. For example, both quantitative and qualitative overload were related to job tension ($r = .4$ and .6 respectively), but their relationships to self-esteem were more specialized. Quantitative overload was related to low self-esteem among administrators ($r = .7$) but not among professors. The self-esteem of professors was related to qualitative overload, however ($r = .3$).

The research at Goddard and Kennedy (French and Caplan, 1973) replicated the findings on the prevalence of overload and on the associated symptoms of psychological strain; it also investigated indicators of physiological strain. In general, overload in these work settings is common; 73 percent of the people interviewed at Goddard reported overload, compared to 44 percent of male white collar workers in the nation as a whole. Among the Goddard scientists, when this overload is qualitative it tends to be associated with low self-esteem (r = .3). Like professors, their self-esteem is not threatened by mere quantity, although they show other signs of strain in the presence of quantitative overload.

All this research bears on relationships among factors in the objective environment, the subjective environment, and responses of the individual. Less has been done on explicit criteria of health and illness. Some data are available, however, on physiological responses to overload (Caplan 1971). Twenty-two men at NASA were observed at work for periods of two to three hours on three days. Their heart rates were recorded telemetrically during the same periods, and their cholesterol levels were determined daily. Findings indicated agreement between reported overload and observed numbers of telephone calls and visitors (r = .6). Both subjective and objective overload were related to heart rate (r = .7 and .4) and to cholesterol (r = .4).

In short, we find that conflict often occurs as overload, that the fact and experience of overload is common in the work situation, that qualitative and quantitative overload must be differentiated, and that these forms of overload are clearly associated with symptoms of psychological and physiological strain.

Role Ambiguity

Role ambiguity is conceived as the discrepancy between the amount of information a person has and the amount he requires to perform his role adequately. When we asked people about such matters in the national survey (Study #2), 35 percent said they were unclear about their responsibilities and were disturbed by that fact, and about equal numbers expressed similar feelings of unclarity and disturbance about what their co-workers expected of them, how their supervisors evaluated them, and what their opportunities for advancement might be. Moreover, the experience of ambiguity was correlated with job-related tension (r = .5), with job dissatisfaction (r = .3), with a sense of futility (r = .4), and with low self-confidence (r = .3). It was also associated with low trust and low liking for co-workers. These relationships, reasonably enough, were conditioned by individual differences—especially the individual's need for structure.

At Goddard (Caplan, 1971), the findings on ambiguity as related to dissatisfaction with the job were replicated (r = .4), and ambiguity was also found to be related to feelings of job-related threat to mental and physical well-being (r = .4) and to lack of utilization of intellectual skills and knowledge (r = .5). Moreover, the ambiguity experience at Goddard was relatively common; it was reported by 60 percent of the men in that organization.

More recently, at Kennedy Space Flight Center (French and Caplan, 1973), we are attempting to look more deeply at the strains associated with ambiguity and to differentiate forms and aspects of ambiguity. In that study we find role ambiguity to be associated, as before, with dissatisfaction with the job (r = .4), with job-related threat (r = .5), and slightly with low self-esteem (r = .2). But we also find it to be related to anxiety (r = .3) and to a four-item scale of somatic symptoms of depression (r = .5).

Our efforts to differentiate the ambiguity concept are modest thus far, but it seems worth distinguishing between ambiguity about the the present role and ambiguity about one's future prospects. Future ambiguity, as measured in the Kennedy study, explains most of the relationship between the overall ambiguity index and such variables as dissatisfaction with the job, job-related threat, and affective depression.

CONCLUSION

To attempt a concluding statement to this paper is inappropriate in at least two respects: The processes that it describes are still at an early stage. Moreover, even the present story line has only begun with this paper; Dr. Cobb and Dr. French will complete it. In doing so they will draw on the programmatic framework described above and continue the development already sketched—from role stress to role conflict and ambiguity, from conflict to role overload and "underload," from overload in general to quantitative and qualitative overload, and from ambiguity in general to present and future ambiguity. Such differentiation, we are convinced, is as necesary for understanding as for solving the pragmatic problems of job stress.

BIBLIOGRAPHY

Caplan, R.D.: *Organizational Stress and Individual Strain: A Social-Psycho-*

logical Study of Risk Factors in Coronary Heart Disease Among Administrators, Engineers and Scientists. Unpublished doctoral dissertation, University of Michigan, Ann Arbor, 1971.

French, J.R.P., Jr., and Caplan, R.D.: Organizational stress and individual strain. In Morrow, Alfred J. (Ed.): *The Failure of Success.* New York, AMACOM, 1973, pp. 30-66.

French, J.R.P., Jr., Tupper, C.J., and Mueller, E.F.: *Work Load of University Professors.* Cooperative Research Project No. 2171, University of Michigan, 1965.

Kahn, R.L.: Work modules. *Psychology Today,* February, 1973, pp. 35-39, 94-95.

Kahn, R.L., Wolfe, D.M., Quinn, R.P., Snoek, J.D., and Rosenthal, R.A.: *Organizational Stress: Studies in Role Conflict and Ambiguity,* New York, Wiley, 1964

Kraut, A.I.: *A Study of Role Conflicts and Their Relationships to Job Satisfaction, Tension and Performance.* Doctoral dissertation, University of Michigan, Ann Arbor, University Microfilms, 1965, No. 67-8312.

Lazarus, R.S.: *Psychological Stress and the Coping Process,* New York, McGraw, 1966.

Rice, A. K.: *Productivity and Social Organization: The Ahmedabad Experiment.* London, Tavistock Publications, 1958.

Sales, S.M.: *Differences Among Individuals in Affective, Behavioral, Biochemical and Physiological Responses to Variations in Work Load.* Doctoral dissertation, University of Michigan Ann Arbor, University Microfilms, 1969, No. 69-18098.

Selye, H.: *The Stress of Life,* New York, McGraw, 1956.

Trist, E.L., Higgin, G.W., Murray, H., and Pollock, A.B.: *Organizational Choice.* London, Tavistock Publications, 1963.

Trist, E.L., and Bamforth, K.W.: Some social and psychological consequences of the long-wall-methods of coal-getting. *Human Relations, 4:*3-38, 1951.

Work in America. Report of a Special Task Force to the Secretary of Health, Education & Welfare, MIT Press, Cambridge, Massachusetts, and London, England, 1973.

CHAPTER 7

ROLE RESPONSIBILITY:
THE DIFFERENTIATION OF A CONCEPT*

SIDNEY COBB

THE PURPOSE OF THIS presentation is to review a set of diverse data suggesting that there are substantial health costs associated with heavy responsibility. Though Western society has managed to develop a system of rewards capable of inducing people to take responsibility, it has only recently begun to appreciate the personal costs involved. Perhaps this beginning appreciation of the costs contributes to the frequency with which administrative posts in our universities, and especially in our medical schools, are remaining vacant.

In the preceding paper, Dr. Kahn has discussed the gradual differentiation of the general concept of role stress into a variety of categories including work load. He has pointed out the value of discriminating between the quantitative versus the qualitative aspects of work load. Now it is time to turn our attention to responsibility which is often confounded with overload, but is conceptually quite distinct.

Responsibility, when well handled, usually brings substantial rewards in public esteem and money. However, the risks of failure are often great. When the wherewithal to avoid failure is not available to the person who has to accept the blame, the situation is particularly difficult. Persons who have strong consciences and/or moderately high fear of failure would seem to be at excess risk in jobs involving heavy responsibility. However, these are just the persons that are desired in responsible positions because they are the

*The new research here reported was supported in part by grant No. K3-MH-16709 from NIMH and an informal contract with Walter Reed Army Institute of Research.

ones who are likely to perform well. Put another way, those with weak superegos have no problem with responsibility. If they have responsibility they don't take it seriously. But then, they are seldom put into responsible positions.

Responsibility may be divided into that primarily involving persons, their work schedules, their rewards and their futures; and that involving things, i.e., money, equipment and projects. Presently it will be shown that responsibility for persons, particularly for their futures, is the one more likely to produce strain, and it seems reasonable to suppose that the distance—spatial, social and psychological— between the responsible individual and those persons for whom he is responsible might materially mitigate the resulting strain. Certainly Milgram's (1965) experiments would support this point of view. Milgram demonstrated that the greater the distance between the experimental subject and the stooge receiving the electric shocks, the more vigorously the subject was willing to apply shocks to "train" the stooge. In other words the subjects felt it difficult to administer shocks to those with whom they were in close contact.

In considering the risk of responsibility, attention will be focused on peptic ulcer, particularly duodenal ulcer and myocardial infarction, plus the physiological factors that appear to contribute to these diseases. First we will examine a study in which the actual measurement of responsibility was attempted. Then we will turn to several groups for whom taking close personal responsibility for people constitutes a substantial portion of their duties, namely physicians, first line supervisors, fathers, and air traffic controllers.

In a study done in the National Aeronautics and Space Administration by Caplan (1971), there is evidence bearing on the risk factors for coronary heart disease. First of all, he finds that being a smoker is significantly associated with responsibility for persons and with responsibility for things, as measured by two brief self-report measures. Since smoking appears to increase the risk of coronary heart disease, this finding suggests that the risk of a myocardial infarct might be increased by responsibility.

Diastolic blood pressure is another coronary risk factor which is related to one of the responsibility measures. It is positively associated with the amount of time the person says he spends in exercising responsibility for others' futures, but not with other aspects of responsibility. Furthermore, this is only true of persons who are in the upper third in any one of four of the dimensions of the type A personality that were developed by Sales (1969). These four dimensions are called Involved, Striving, Persistence, Leadership, and Positive Attitudes toward Pressure. Each of these measures contains items that could be construed as measuring conscience, as for example, "In general I approach my work much more seriously than most of the people I know." Furthermore, three of the four measures are positively and significantly correlated with the report that the person imposes deadlines on himself.

Cholesterol level is associated with the self-report of time spent exercising responsibility for others' futures, but only in those who are high on the first two type A personality dimensions mentioned above, to wit: Involved Striving and Persistence.

We have, then, modest evidence that responsibility for persons, rather than for things, particularly for people's futures, contributes to coronary disease risk among those who are overly conscientious. Furthermore, there is in Caplan's work ample evidence that this quality of being conscientious is associated with responsibility. Therefore we might say that the conscientious person is encouraged to take on responsibility for people and their futures, is rewarded for it with increased status and income, and pays the price of increased risk of myocardial infarction. This is a rather tenuous chain, and the relationships are not strong, but they certainly suggest that the matter should be studied further.

Obviously, physicians take a great deal of responsibility for people, so they might have a high risk of relevant disease. Unfortunately, this hypothesis is difficult to test, because physicians are set apart from other mortals by the fact that they may well get both better diagnosis and better treatment. This means that biases may be introduced into both morbidity and mortality comparisons with groups outside the profession. This applies particularly to the data on peptic ulcer. The two best studies of the frequency of ulcer disease (Doll and Jones, 1951; Alsted, 1954) have shown a great excess of ulcers among physicians, but the findings are interpreted with caution. However, within the profession, comparisons are more appropriate. It is clear that anesthesiologists and general practitioners take more life and death responsibility than do dermatologists and pathologists. Russek (1960) found that the anesthesiologists and general practitioners reported two to three times as much coronary heart disease as the dermatologists and pathologists. Without comparable incidence and/or mortality information, such prevalent differences may be misleading. However, according to a British study of physicians (Morris, *et al.* 1952) the incidence of, and mortality from, myocardial infarcts is about twice as high in general practitioners aged forty to sixty, as in other members of the profession of comparable age. Again, we are left only partially convinced.

Let us turn to first line supervisors. Foremen have been repeatedly reported to have an excess of peptic ulcer. Doll and Jones (1951) have done an important study of the frequency of ulcer disease by occupational groups. They found that foremen and executives combined had more duodenal ulcer than expected, while various skilled and unskilled occupations had frequencies approximating the expected. In a large electrical company in Holland, Vertin (1954) found that foremen and assistant foremen had significantly more peptic ulcer than the workers they supervised. Likewise, Pflanz, *et al.* (1956) and Gosling (1958) have reported similar findings. The above studies were all done in Europe. That the same is true in this country is illustrated by the data in Table 7-I which I collected some years ago

TABLE 7-I

PREVALENCE OF PEPTIC ULCERS IN FOREMEN AND CRAFTSMEN WHOM THEY SUPERVISE

Certainty of Ulcer Dx.	Prevalence Per 100		Approximate Ratio Of Prevalences
	Craftsmen	Foremen	
Treated	10	12	1
X-Ray Confirmed	7	15	2
Gastrectomy	1	12	10
Overall	18	38	2
No. of Men	272	26	

CHI Squared = 17.6, p<0.001.

(1963). This group of foremen obviously has more severe peptic ulcer than the men they supervise. I don't know of any studies which specifically and separately identify foremen and deny that they have excess peptic ulcer.

Aside from this matter of immediate and close responsibility for people on the job, Vertin has pointed out that parental responsibility, as measured by the number of children, is strongly related to the proportion of men who have been treated for peptic ulcer. Unfortunately, Vertin has not controlled for age and responsibility on the job, nor are his population estimates fully appropriate to his estimates of the number of ulcer cases. Nevertheless, his results are very suggestive of a substantial relationship. This topic deserves further investigation.

I am not acquainted with any adequate morbidity data on coronary disease, hypertension or diabetes among foremen. The available mortality data are suspect because of problems with the occupational classification that doesn't clearly separate the first line supervisor from the supervisee. With the realization that we need more and better information about the health of foremen, it is time to turn to one other group.

Everyone who has ever flown in a commercial airliner either has or should have prayerfully hoped that his plane is being guided by a series of conscientious and responsible air traffic controllers. Those who have read *Airport* (1968) are aware of the terrible guilt that can arise from responsibility for an accident. I cannot imagine an occupational situation in which an individual could be more directly responsible for the lives of large numbers of people with whom he is constantly in voice contact.

With these thoughts in mind, it is appropriate to examine some data that I have recently gathered. They will be presented in greater detail elsewhere (Cobb & Rose, 1972). The findings are summarized in Table 7-II. The study involved the 1969-70 licensing renewal

TABLE 7-II

SUMMARY OF THE EVIDENCE LINKING THE DEMANDS OF THE JOB
AS A JOURNEYMAN AIR TRAFFIC CONTROLLER IN A TOWER
OR CENTER WITH CERTAIN DISEASES

	Hypertension	Diabetes	Peptic Ulcer
1. Point Prevalence	(++) 4	++ 2.3	++2
2. Annual Incidence	++ 5.6	+2	+1.5
3. Age Effect	++	+	+
4. Traffic Density Effect	++	NI	++

++ Evidence is strong and significant
+ Results in the preducted direction but numbers are too small for significance
NI—No Information
(   ) Bias may account for a portion of the effect.
The numbers indicate the factor by which the frequency among ATC's exceeds that in 2nd Class Airmen.

examinations for 4,325 male journeymen air traffic controllers working in towers or centers. These were compared with the same examinations for 8,435 male second class airmen also applying for renewal of their licenses. Though the examinations were the same and were conducted by the same set of physicians, it is quite possible that differences in prevalence between the two groups could be due to difference in licensing practices and/or differences in the probability of a person reapplying, given that he knows that he has a disease which might prevent the renewal of his license. This problem has been faced in a number of different ways as will be seen from the data to be reviewed. In all of the comparisons age is taken into account by making age specific comparisons and/or age adjustments.

In Table 7-II we are concerned with hypertension, diabetes and peptic ulcer. These are the only diseases from the list of those known to be responsive to stresses in the social environment which are both separately identified and sufficiently frequent for analysis. For example, migraine headaches were too infrequent and rheumatoid arthritis was not separable from the other arthritides in the coding system used by the Federal Aviation Administration.

The prevalence data are presented on the first line of the table. Hypertension is four times as prevalent among air traffic controllers as among second class airmen, after adjustment for age. This is a highly significant difference, but it appears to be contaminated by the fact that given a diagnosis of hypertension, a second class airman is slightly more likely to be refused a license. It is, however, quite improbable that the small licensing bias accounts for the large difference in prevalence.

A more serious licensing bias exists with regard to those who have severe diabetes. As a result the analysis is confined to diabetes controlled by diet alone, for which the licensing bias is trivial. The prevalence of mild diabetes controlled by diet is 2.3 times as high for air traffic controllers as it is for second class airmen ($P < 0.05$).

Peptic ulcer is unfortunately not separable into gastric and duodenal, but it is reasonable to assume in this relatively young population that gastric ulcer constitutes only a small fraction of the total (Doll & Jones, 1951; Eusterman, 1947). Fortunately there is no significant licensing bias for peptic ulcer. The disease is twice as common among air traffic controllers as among second class airmen (P < 0.001).

The incidence data (number of new onsets per unit of population per year) on line 2 of the table presents a confirmatory pattern. Since the numbers are smaller, statistical significance is not achieved for diabetes controlled by diet alone, nor for peptic ulcer.

Line 3 summarizes the data on differential age of onset. The age at which each of these diseases began was younger for the controllers than for the airmen. For hypertension the means were forty-one and forty-eight years of age respectively (P < 0.025). For the others the differences were not statistically significant.

On line 4 is presented a comparison within the group of air traffic controllers. The men were divided according to the traffic density at the tower or center at which they were employed. The prediction was that those working under high traffic density would have greater responsibility and greater work load, hence a greater point prevalence of illness. This proved significantly true for hypertension (P < 0.025) and for peptic ulcer (P < 0.001). The number of cases of diabetes controlled by diet was too small to examine for the effect of traffic density since the comparison has to be made on an age specific basis.

For obvious reasons we were interested in coronary disease, but unhappily the licensing bias was too great to draw any inferences. Since hypertension and diabetes appear to be increased in air traffic controllers, and since serum phospholipids levels and the frequency of ECG abnormalities (Hale, *et al.* 1971) are reported to be elevated, it will not be surprising if a proper cohort study should reveal excess incidence of myocardial infarction.

In summary, it can be said that there is scattered evidence that diabetes, hypertension and myocardial infarction as well as peptic (presumably duodenal) ulcer are unduly common among persons subject to heavy, close personal responsibility for the lives of other people. The evidence is not fully consistent, but then all the evidence is not in. The theory says that conscientious people get selected for responsible positions. Since both status and money go with the positions, these people accept the responsibility. The evidence with regard to air traffic controllers is the most striking, perhaps because the situation for them is the most extreme. Further investigation will be required before we can be sure that responsibility is the central problem because it is commonly associated with work overload, which may well contribute in its own right to the health problems noted.

BIBLIOGRAPHY

Alsted, G.: Social and public health aspect of peptic ulcer. *Gastroenterology, 26*: 268, Feb., 1954.

Caplan, R.D.: *Organizational Stress and Individual Strain. A Social-Psychological Study of Risk Factors in Coronary Heart Disease among Administrators, Engineers and Scientists.* Doctoral dissertation, University of Michigan, Ann Arbor, 1971.

Cobb, S., French, J.R.P., Jr., and Mann, F.C.: An environmental approach to mental health. *Ann NY Acad Sci, 107*:596, 1963.

Cobb, S., and Rose, R.M.: Hypertension, peptic ulcer and diabetes in air traffic controllers. *JAMA,* In press.

Doll, R., and Jones, A.F.: *Occupational Factors in the Aetiology of Gastric and Duodenal Ulcers,* Medical Research Council Special Report Series, No. 276, London, HMSO, 1951.

Eusterman, C.B.: Newer phases of gastroduodenal ulcer. *Gastroenterology, 8*:755, 1947.

Gosling, R.H.: Peptic ulcer and mental disorder. *J Psychosom Res, 2*:285, 1958.

Hailey, A.: *Airport,* New York, Doubleday, 1968.

Hale, H.B., *et al.*: Neuroendocrine and metabolic responses to intermittent night shift work. *Aerosp Med, 42*:156-162, February, 1971.

Melton, C.E., *et al.*: *Physiological responses in air traffic control personnel:* O'Hare Tower, FAA-AM-71-2, Department of Transportation, Washington, D.C., 1971.

Milgram, S.: Some conditions of obedience and disobedience to authority. *Human Relations, 18*:57, 1965.

Morris, J.N., Heady, J.A., and Barley, R.G.: Coronary heart disease in medical practitioners. *Br Med J, 1*:503, 1952.

Pflanz, M., Rosenstein, E., and von Uexkull, T.: Socio-psychological aspects of peptic ulcer. *J Psychosom Res, 1*:68, 1956.

Russek, H.I.: Emotional stress and coronary heart disease in American physicians. *Am J Med Sci, 240*: 711, 1960.

Sales, S.M.: *Differences Among Individuals in Affective, Behavioral, Biochemical and Physiological Responses to Variations in Work Load.* Doctoral dissertation, University of Michigan, Ann Arbor, 1969.

Vertin, P.G.:*Bedrijfsgeneeskundige Aspecten van het Ulcus Pepticum.* Thesis, Groningen, Hermes, Eindhoven, 1954.

CHAPTER 8

PERSON ROLE FIT

JOHN R. P. FRENCH, JR.

THE PURPOSE OF my paper is to refine and to qualify the findings on the main effects of job stress on individual strain. Both Dr. Kahn and Dr. Cobb have already reported that the effects of organizational stresses on strain within the individual will vary depending upon the personality of the individual. I will focus on two additional conditioning variables: first the goodness of fit between the environment and the person; and second, the conditioning effects of social support.

First let me sketch very briefly a theory about person-environment fit, developed by French, Cobb and Rodgers. One kind of fit between a man and his job environment is the degree to which his skills and abilities match the demands and requirements of the job. Another type of fit is the degree to which the needs of the man are supplied in his job environment, for example, the extent to which his need to utilize his best and highest abilities is satisfied by his current job. Our basic assumption is that both forms of misfit will cause job dissatisfaction, depression, physiological strains, and other symptoms of poor mental health. In order to be able to test this theory, we need to be able to measure quantitatively the goodness of fit between the man and the job. This we have done by asking the man to rate the quality of his job environment along a quantitative scale, for example, "the responsibility you have for the work of others." On the same scale the person was then asked to rate "the responsibility you would *like to have* for the work of others." Then we derived a quantitative score of the goodness of fit, by subtracting the actual score for the job environment from the desired score on the same dimension of the job environment. A given individual might want

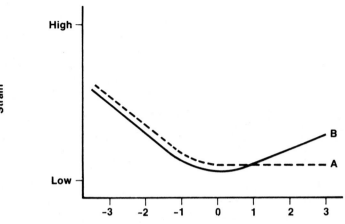

Figure 8-1. Two hypothetical P-E fit curves.

more responsibility, or less responsibility. Or there might be a perfect fit, with no discrepancy between what he has and what he wants. Figure 8-1 illustrates such a scale measuring the person-environment fit. The negative numbers represent a deficiency—the person wants more of the environmental variable than he has. The zero represents a perfect fit, and the positive numbers represent an excess. One hypothesis suggests that an increasing deficiency in supplies to meet a person's needs will lead to increasing strain, as shown in Curve A. According to this hypothesis, an excess of supplies to meet the need will make no difference. For example, a man who has completely satisfied his hunger will not be more satisfied with more food. The man will still show the same low level of strain as if he had a perfect fit. Another hypothesis, shown in Curve B, assumes that an excess may also result in increased strain. This seems reasonable, for example, when we think about responsibility for other people. Too much responsibility may be more than a person can bear easily, whereas too little responsibility may mean that he has a low level, uninteresting job. Both hypotheses seem reasonable, and we might expect Curve A for one type of environmental variable and Curve B for another type. In either case, our theory of person-environment fit should account for additional variance in mental health over and above the variance we can account for by the linear effects of job stresses and of personality variables.

Now let us look at some tests of these theories. I would like to report findings from two different studies, one conducted at Goddard Space Flight Center and the other at Kennedy Space Center. In both studies, we measured the actual state of the job environment along about a dozen different dimensions. Then we measured the desired state of the environment along those same dimensions. The measures of both the actual and the desired state were constructed from several questionnaire items. Thus we could derive quantitative measures of the goodness of fit between the man and his job environment.

Figure 8-2 shows the results for a sample of managers, engineers, and scientists at Goddard Space Flight Center. We have plotted job satisfaction against person-environment fit, so we should expect the highest satisfaction at zero, where there is a perfect fit between what a man has in his job environment and what he wants.

Looking first at the curve for goodness of fit with respect to role

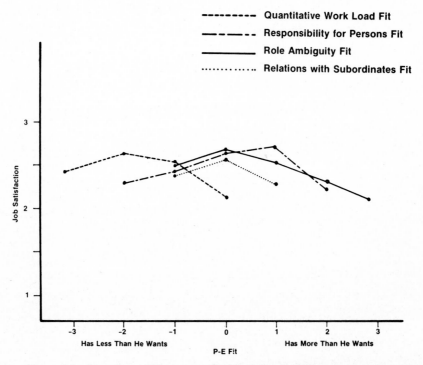

Figure 8-2. Data from 206 men at Goddard Space Flight Center. The **N**'s at the ends of the curves are typically from 4 to 13.

ambiguity, we see that the curve is displaced to the right, indicating that most people want less role ambiguity than they actually have. Given this distribution, the curve shows the expected inverted "U" shape with the highest satisfaction where there is a perfect match between the amount of role ambiguity experienced and the amount desired; for those who want either less role ambiguity or more, there is lower job satisfaction.

The measure of relations with subordinates deals primarily with quantity and quality of work they expect from their superior. As we can see from the curve, there is generally good fit between the perceived expectations and the desired expectations. As predicted, job satisfaction is highest when the fit is perfect, and falls off with either a positive or a negative discrepancy.

The curve for goodness of fit with respect to the responsibility for persons shows the generally expected shape, but the maximum satisfaction does not occur among men with a perfect fit, but instead among men who have slightly more responsibility for others than they would like to have. Other data show that this same group had the lowest level of cholesterol. Perhaps the stressful effects of a slight excess in responsibility for other people is offset by a better fit with respect to utilization of one's best abilities and participation in decision-making.

Finally, Figure 8-2 shows that these men generally want more quantitative work load than they have, a rather surprising fact, since the reported levels of work load at Goddard are higher than in our national samples. Again, the curve shows the expected shape, but now there is a large displacement to the left. It is the eager beavers who show the highest job satisfaction. Perhaps the direction of causation is reversed here: high satisfaction with work causes a person to want more of it.

There were similar significant relations between job satisfaction and person-environment fit with respect to five other dimensions of the work environment: namely, participation, responsibility for things, qualitative work load, the utilization of abilities, and the opportunities for advancement. In sum, nine out of eleven dimensions of job stress showed significant relations between person-environment fit and job satisfaction.

The same dimensions of person-environment fit were correlated with a measure of job-related threat which probed the extent to which various kinds of job stress posed a threat to perceived health and well-being. The results were similar to the findings we have just reported for job satisfaction. For eight out of eleven dimensions, poor fit is associated with high job-related threat. Frequently, the relationship is curvilinear, conforming to one or the other of the theoretically predicted curves.

Now let us turn to the findings from Kennedy Space Center. In this study, we developed measures of person-environment fit with respect to six job stresses: role ambiguity, subjective work load, participation, responsibility for things, responsibility for people, and underload. We correlated these measures of goodness of fit with five measures of psychological strain, including job satisfaction, job-related threat, anxiety, work-related depression, and somatic depression. The two measures of depression were factors derived from the Zung scale for depression. Sixteen out of the thirty measures of curvilinear correlation were significant. When these same measures were administered six months later, there was a good replication of the findings that person-environment fit is significantly related to psychological strain.

Of the sixteen significant correlations between goodness of fit and psychological strain on the first questionnaire, twelve were linear and four were significantly curvilinear. These four are shown in Figure 8-3. Again we notice that most men want less role ambiguity than they have, and we can now add that most men would like more participation than they have. As predicted, all four curves showed the lowest strain where person-environment fit is perfect. As in Figure 8-2, the number of cases at the extreme ends of the curves is quite small, but nevertheless, there is a significant departure from a linear relation.

Our measures of person-environment fit are only rarely related to physiological measures, and then the correlations are quite low and inconsistent. However, we have found significant correlations with cholesterol, systolic blood pressure, and glucose. All of these physiological measures are related to person-environment fit in the expected direction.

We may summarize our findings on person-environment fit by

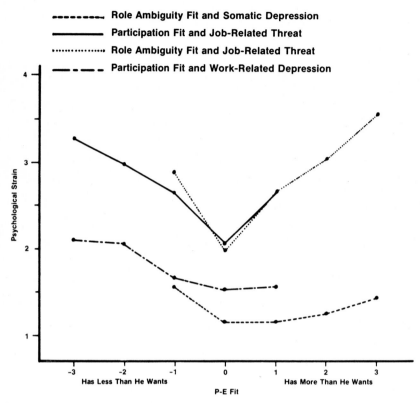

Figure 8-3. Data from 165 men at Kennedy Space Center.

saying that many dimensions of fit are related quite strongly and in the predicted direction to a variety of measures of psychological strain. Many of these relations between fit and psychological strain show the forms of curvilinearity predicted by the theory. Only a few of our measures of fit are related, but then quite weakly and in the predicted direction, to our measures of physiological strain. Our tests of the hypotheses tend to be weak because: (1) we have too few cases where there is a strong discrepancy between what a person has and what he wants; (2) the measures of fit, like other measures of discrepancy scores, tend to be unreliable; (3) the measures are probably biased by wishful thinking and dissonance reduction; and (4) there are some unsolved problems of analysis. Considering the methodological difficulties, the empirical results we have obtained show about as strong support for our theory of person-environment fit as we could expect. An unequivocal test of the theory will require an improvement in methods.

Now let us look at the conditioning effects of social support on the relationship between job stress and individual strain. Our conceptions and our measures of good, supportive relationships are taken from the work of Rensis Likert and others at the Institute for Social Research. The measure on good relationships with the immediate supervisor included items on "the extent to which your superior is willing to listen to your problems," "the extent to which your superior has confidence in you and trusts you," "the extent to which you can trust your superior and have confidence in him." The items measuring relationships to the work group included, "the extent to which persons in your work group are friendly and easy to approach," and "the extent to which persons in your work group are willing to listen to your problems," and "the degree of cooperation in the group." The measure of relations with subordinates included similar items, and also items on mutual expectations with regard to the quantity and quality of work.

Table 8-I shows how the effect of role ambiguity on serum cortisol

TABLE 8-I

THE EFFECT OF RELATIONS WITH SUBORDINATES ON THE
CORRELATION BETWEEN ROLE AMBIGUITY AND SERUM CORTISOL

Relations with Subordinates	r	$p <$	n
Poor	.35[a]	.01	60
Medium	.26	.05	57
Good	.06[a]	n.s.	76

[a] The difference between these two correlations is significant at $p < .05$.

varies with relations with subordinates. For men who have poor relationships with their subordinates, there is a significant positive correlation between role ambiguity and cortisol; but for men who have good relations, there is no such correlation. This pattern of findings is not obtained when relations with one's immediate superior is used as a conditioning variable.

Table 8-II shows that the effects of quantitative work load on several physiological variables varies depending on the quality of relationships with the superior, the work group, and subordinates. Diastolic blood pressure is related to work load, but only among those

TABLE 8-II

THE EFFECT OF RELATIONS WITH OTHERS ON THE
CORRELATION BETWEEN WORK LOAD AND
PHYSIOLOGICAL STRAIN

Relations with Strain	Quality of Relations		
	Poor	Medium	Good
Immediate Superior			
Diastolic blood pressure	.33[a]	—.06	.06
Work Group			
Serum glucose	.31[a]	.03	—.01
Subordinates			
Systolic blood pressure	.24	.22	—.09
Diastolic blood pressure	.31[a]	.19	—.13
Serum glucose	.36[a]	.06	.04

[a] $p < .01$. The correlations for those with poor relations are always significantly higher than the correlations for those with good relations at $p < .05$.

who have poor relationships with their immediate superiors. Serum glucose is related to work load, but again only among those who have poor relationships with their work groups. Relations with subordinates have even stronger conditioning effects on systolic blood pressure, on diastolic blood pressure, and on serum glucose. Thus good relations with others, especially with one's subordinates, serve as a buffer between the stress of quantitative work load and the physiological strains which may result. We do not obtain similar findings when we use other measures of work load or other measures of physiological strain. More surprising, supportive relations with others do not seem to condition the effects of job stress on psychological strain.

Table 8-III shows that the quality of relations with one's superior does condition the effects of work load on smoking. For half a dozen

TABLE 8-III

THE EFFECT OF RELATIONS WITH THE IMMEDIATE SUPERIOR
ON THE CORRELATION BETWEEN SUBJECTIVE WORK LOAD
AND NUMBER OF CIGARETTES SMOKED[a]

Subjective Work Load Measure	Relations with Superior		$r$poor $r$good
	Poor N=24	Good N=28	$p <$
Number per week of:			
Outgoing phone calls	.36	—.26	.025
Incoming phone calls	.32	—.06	.10
Office visits	.06	—.26	n.s.
Self-initiated meetings	.30	—.34	.025
Other-initiated meetings	.09	—.44[b]	.05
Total activities	.36	—.17	.05

[a] Smokers only.
[b] $p < .05$ (two-tailed).

different measures of work load, such as number of phone calls, number of office visits, and number of meetings, there is a positive correlation between work load and smoking among those men who have poor relations with their superior, whereas there is a negative correlation among those men who have good relations with their superior. When we use relationships with the work group or with subordinates as the conditioning variable, we find similar but weaker relations.

Finally, we find that the amount of stress reported coming from other parts of the organization, for example, the amount of stress coming from other branches, is positively related to pulse rate, but again only in those men who have poor relations with their subordinates. The relations with one's superior and with one's work group have similar conditioning effects on the effect of stress from other parts of the organization on pulse rate. In all cases, good relationships act as a buffer between stress and strain.

To summarize, in many cases, we have found that the effect of job stresses on individual strain are eliminated by the buffering effect of good supportive relations with one's superior, one's work group, and one's subordinates. However, in many other cases, especially when we are dealing with psychological strain, we do not find such buffering effects. Additional analyses of the effect of supportive relationships at Kennedy Space Center are now in progress, and it is hoped that they may further clarify some of these relationships. However, it will probably be necessary to do additional studies in which we examine in more detail the *processes* by which supportive relationships prevent strain. Perhaps social support reduces the objective environmental stresses, or perhaps it only affects the perceived stresses, or perhaps it enables the person to cope more effectively with stresses which are there. These and other hypotheses require further research.

Now I would like to end by stating six general conclusions which I think can be drawn from the research reported by Dr. Kahn, Dr. Cobb and myself.

1. Both our theoretical thinking and our empirical findings have led us to develop more differentiated concepts and measures for describing role stress. We have progressed, for example, from talking about a fairly global role conflict to distinguishing between conflict and overload; then we distinguished between quantitative and

qualitative overload. Next, we added responsibility for persons and responsibility for things, as more specialized aspects of the work load. Then we added underload as an additional role stress discovered at NASA. To date we have examined the effects of fourteen objectively measured job stresses and twenty subjectively measured job stresses.

2. As we have become more differentiated in our thinking about job stress, we have also improved the reliability and validity of our measures, not only by developing different measures for the different stresses, but also by improving the measures by adding items, by eliminating unsatisfactory items, by decreasing response biases, etc.

3. These more differentiated and improved measures have paid off; they have yielded new findings which were previously masked. In general, our findings support the specificity hypothesis that specific kinds of job stress interacting with specific personality characteristics result in specific strains in the person, including psychological strains, physiological strains, and psychosomatic diseases.

4. The simple main effects of various role stresses have been further qualified by discovering many conditioning variables which permit us to explain more of the variance in strain. In addition to the conditioning effects of personality variables which have been described by Drs. Kahn and Cobb, we have also found that the goodness of fit between the characteristics of the person and the characteristics of his job will influence the effects of job stresses. Finally, we are beginning to find that supportive relations with other people can often act as an effective buffer between job stress and strain, particularly physiological strain, within the person.

5. As our work has progressed, we have added new measures of psychological strains, of physiological strains, and of disease entities. These new measures have revealed ever more ubiquitous consequences of job stresses.

6. As we develop more refined and more empirically validated knowledge of the nature of role stress and their effects, we will be better able to design intervention programs to prevent their undesired effects on mental and physical health. Currently we are in the middle of our first study of the feasibility of reducing the risk of coronary heart disease by altering role stresses. We hope that the kind of findings presented here can be applied in additional experiments on the prevention of strain.

CHAPTER 9

HEALTH STATUS ASSESSMENT:

An Untapped Source of Management Information

DAVID R. SCHWARZ

WHAT WOULD HAPPEN if a medical director reported to his chief executive officer that three times the normal incidence of increased blood pressure had developed in a certain department in a given period; and that evidence indicates that this trend precedes increases in heart attacks by about six months? How often does the medical director speak with the chief executive officer? How many companies use available medical expertise as an integral part of management information to guide and support better organization performance?

These questions were posed in a subgroup dealing with how organizational structure, climate and function may influence occupational stress* reactions. The conclusion reached was that a nearly universal communication gap between line managers† and occupational health practitioners impedes the pooling of skills needed to deal with such questions. An open circuit exists in the feedback loop

---

*"Stress" is used here as defined by Lazarus in chapter 12.

Participants in the subgroup discussion which generated this chapter were: Caesar Briefer, M. D.; John R. P. French, Jr., Ph.D.; Prof. Leopold Gruenfeld; Howard Hess, M. D.; R. Edward Huffman, M. D.; Thelma Larsh; Cavin P. Leeman, M. D.; William Liggett; Bruce Margolis, Ph.D.; Leo Miller; Robert Miller; Eileen Morley, Ph.D.; Charles R. Pfenning; Granville Walker, M. D.

†For the purpose of this discussion the term "line manager" is any person in the chain of command directly concerned with production of the products or services furnished by the organization to its consumers and who has responsibility for the supervision of other employees and the economic utilization of material resources. The chief executive officer is the person at the apex of this chain of command.

to decision making that prevents the constructive use of employees' physical and mental health data to improve the effectiveness of general management.

Increasing evidence shows that how work is organized, the climate created by management, and the clarity with which responsibilities are defined can produce physical and emotional symptoms which adversely affect individual and organizational performance.

Managers frequently discuss how to improve communications and how to reduce foul-ups in carefully developed operating plans which cause low productivity and disappointing financial outcomes. Although they often recognize "people problems," they rarely consult the health staff, frequently the first to observe and deal with personal dysfunction. Nor does the medical department initiate action and offer to contribute to the resolution of these problems. It is generally content to remain in its traditional therapeutic role. Why is this so? Would it be worthwhile to try to change this condition?

One problem is that many health specialists now operating in organizations do not generally feel comfortable dealing with management problems. Moreover, they don't think managers are eager to have them do so. Physicians, psychiatrists, counselors and personnel experts dealing with stress in organizations currently see their principal role as performing therapeutic services for individual employees. Neither their job descriptions, their relations within the organization, nor their professional orientations encourage them to try to bring about organizational change. Many believe that most managers view these therapeutic activities as a desirable fringe benefit for employees, but would resent attempts to apply them to improving management. The feeling that this resentment might endanger the effectiveness of the therapist in his mutually accepted role further discourages efforts in this direction.

Managers in general do not regard the incidence of unfavorable stress reactions in relation to their effectiveness as managers, nor does stress appear to them as an impediment to achieving organizational goals. On the contrary, some observers suggest that the ability to overcome (not reduce) stress is considered a "badge of honor" among managers.

Unquestionably, the medical, personnel and counseling staff in work organizations are favorably positioned to spot early symptoms of stress among employees. The direct experience reported at the conference supports the conviction that a competent medical and counseling service often can identify and successfully treat physical and mental reactions to stress in individual cases. Where the stress involves relations with the immediate supervisor or other employees, the practitioner may be brought into a training or consultative role which can be very effective. When the stress is rooted in a more general management situation as indicated, for example, by the clustering of functional or psychosomatic complaints in a particular department, initiating corrective action begins to get "sticky." This "stickiness" is influenced by how well the manager and therapist know each other as well as their hierarchically defined roles. It is also related to the manager's position in the formal structure, how he is affected by the organizational climate, and the clarity with which his responsibilities are defined. In other words, the system becomes increasingly non-accepting of intervention by the therapists as the problem moves from individual distress to systematic malfunction.

Data on trends which signal impairment of performance could be developed into sensitive diagnostic and predictive indices of management effectiveness. Collecting and analyzing these data is appropriate to the professional skills of the health services staff, but developing how this information should be used for decision making involves a sharing of insights, techniques and purposes with line managers. Few formal channels are available to accomplish this in most organizations.

Finally, most occupational health specialists feel unqualified to deal with the professional skills and specialized vocabulary of management. They naturally feel more comfortable in roles for which they were formally trained. Also many of them are not sure that enough knowledge is available about cause and effect so that closer collaboration with managers could lead to beneficial changes.

Line managers are trained to think and operate in terms of profits, growth, budgets, control and chain of command. Making the "bottom line" result acceptable has to be their major preoccupation. They are concerned with action: getting things done through others who report to them. Judgments must be made about cost effectiveness under conditions where there is rarely enough hard data available to be sure of consequences. The pressure of meeting the budget is constant.

Although most managers consciously want to be humane, and think of themselves as such, they generally take pride in a "hard-nosed" approach to what makes the organization and their own careers prosper. The simple dictum is: "No dough; no go!" Among the classic functions of managing: planning, organization, control and evaluation, most of the energy goes into control. A manager who is unable to keep costs and expenses in line doesn't last very long. Salaries and wages are either a cost or an expense, so these are the first categories in which people are viewed, whatever may be verbalized about people being the organizations most important asset.

If you ask the chief executive officers what the medical department does, the gist of most answers is likely to be: "They treat sick employees." If you ask why the company supports this expense, you will probably be told, "A good company takes care of its employees. We think a good medical service helps us get and keep competent people. Besides, prompt and efficient care can cut lost time." But since health services are ancillary activities the cost effetciveness of which is difficult to measure, budgeting for them is likely to get fairly close scrutiny, especially when money is tight. Not many chief executive officers are aware that statistics about general health might be used diagnostically to locate ineffective organization performance. Even for those who may appreciate this potential, developing a method and measuring results may seem a nebulous, and therefore unattractive proposition.

Although most line managers reflect a general respect for the medical profession, they probably believe that these skills are for healing people, not organizations. Furthermore, many persons, managers included, develop from childhood mixed feelings of dependency and apprehension toward doctors which conflict with the sense of power a manager who is "on top of his job" likes to have. These often deeply ingrained attitudes restrain the manager's seeking out the medical expert as a managerial colleague.

The net effect of the preoccupations and preconditioning of managers and health specialists tends to establish a "comfort zone" relationship within organizations. Functions are separate and the health staff is essentially excluded from participation in management decisions which affect the organization as a whole.

The attitudes of employees, their families and their unions are a third crucial factor in the effectiveness of health services and their potential use for improving work organizations. Acceptance of various forms of health insurance, group practice, third party payments, and institutional medical services is now so widely established in the general population that a good company-supported medical department should find little resistance to its therapeutic role among employees. In fact, improved medical benefits have, for some time, been an issue in collective bargaining.

Many unions have also established their own medical and counseling services. Interest in preventive applications of medical knowledge has been developing more slowly. Currently, there appears to be a marked increase of concern about occupational health factors and the right of employees, often through their unions, to participate in decisions which affect physical hazards in the work environment.

Unions have traditionally concerned themselves with economic and physical conditions of work and are only beginning to consider the implications of organizational structure, climate and function as producers of undesirable job stress. These are areas which, in the past, were conceded to be management's prerogatives. A growing restiveness is evident among employees at all levels, manifested by lack of satisfaction from work which produces physical, emotional and behavioral dysfunction. Labor leaders are recognizing the need to deal with these issues.

It may be that pressure from outside management structure, from unions, government, and the general community, in the form of legislation or other avenues of public action, will be the primary driving force for change. Witness the history of improvement in wages and hours and in occupational safety, culminating in the most recent Occupational Safety and Health Act of 1970, the establishment of a National Institute of Occupational Safety and Health and the special task force report, *Work in America,* issued by the Secretary of Health,

Education and Welfare.* There is no doubt that external forces can exert powerful influence on decision making.

Although what is happening in society as a whole must be taken into account, the primary concern of institutional decision-making must be what is going on within the organization. It is within these boundaries that action for change can best achieve measurable results. The discussion reported here therefore focused largely on internal considerations. To do this, it was helpful to view an organization as a power system, because the distribution of power was identified as a key determinant of the flow of information and response.

If power is defined as "ability to influence decisions," having it or lacking it can be a major factor in producing stress. Six forms of power in organizations were characterized:

1. *Former or hierarchical* power which stems from the established structure and generally follows line management chain of command.
2. *Coercive power* ⎰ which are usually attached to formal power;
   and ⎱
3. *Reward power* ⎰ in the vernacular: "hire and fire power."
4. *Expert power* stemming from special technical competence, not generally shared, which commands respect; e.g., the healing expertise of the physician.
5. *Referent power* which arises from respect for the person; e.g., "I like Jack. He has good judgment and cares about what is happening here."
6. *Gate keeper power*—the ability to control the flow of information; e.g., secretaries setting appointment schedules.

Outside of their own department, medical directors and other health specialists generally have little power in forms 1, 2 and 3. But they should have substantial expert power and are often well placed to develop significant referent power.

Change in organization structure and function generally requires action through the formal power system involving line management and beginning at the top. The climate can be influenced to some degree by expert and referent power. If health practitioners are to

---

*The Secretary's (H.E.W.) Task Force on Work in America: *Work in America*. Boston, MIT Pr, 1973.

use these types of power effectively, they must feed back information and advice in a form consistent with the power concepts of line managers.

Particularly in large, complex organizations, the flow of hierarchical power defines the structure as pyramidal with increasing separation of functions down the chain of command. This type of system, which appears attractive from the viewpoint of control, inherently creates serious communication problems. From the top down, messages become increasingly diffused. Response messages get choked off as they travel toward the top. The functional separation of production, marketing, finance, and research and development reduces opportunities for cross-linking at intermediate levels where interfunctional collaboration might facilitate smooth operation. The pathway for messages across functional lines is often lengthened by chain of command requirements resulting in poor signal-to-noise ratio. In multidivisional companies, communication among the individual profit centers is even further attenuated. In many cases, staff functions, including health services, have secondary relationships with the main power and communication flow which make it difficult to influence action outside the prescribed area of staff expertise.

Inevitably, informal networks* of personal relationships develop within organizations as a necessary means of getting a day's work done against the stiffness of the formal structure. As good manager is expected to know whom he can call in another department (not necessarily the boss) to help him unsnarl a difficulty, even though the paperwork, when it finally comes through, may have navigated all the required channels. It is through such networks that expert and referent power are often most effectively utilized. Much supportive counseling may be accomplished through these interactions. The cumulative effect in facilitating the overall operation may be considerable, but direct impact on decision making and the hierarchical power structure is difficult to trace and identify.

Elementary theory teaches that self-regenerative systems require built-in means for processing corrective feedback signals capable of adjusting the activities of the system to maintain its objectives. In

---

*For a provocative exposition on network approaches to management, see: Donald A. Schon: *Beyond the Stable State*, New York, Random House, 1971.

the typical large business organization, there is a definite limit to the number of signals which can filter back to affect the decision-making process. Most attention is given to relatively simple messages designed into the system to provide quantitative information relevant to the fiscal aims of the organization: sales, cost of goods, gross margins, expenses, gross and net profit, cash flow, capital needs, etc. Feedback about people is most commonly in the form of statistics on number employed, wages and salaries, absentee rates, turnover, lost-time accidents, and the like. At the distance of central control, processing of these latter signals may provide some inferences about how people are performing at the periphery, but this deduced information is not likely to have the same simplicity and force as the basic numerical and financial messages. Instructions for corrective action, or at least some hard questions, will be generated from headquarters within days or weeks after receipt of out-of-control fiscal information. It may be months or years before the system catches up with a condition which is producing substantial over-stress among employees.

Individual responses to work and interactions among persons in the work place involve non-rational dynamic processes which are not readily reduced to rational analysis yielding simple signals. This seems to be one of the primary reasons why "people problems" present such difficulties for highly rationalized management systems. Ultimate corrective action for improving the human environment requires respect for, and sensitivity to, the needs and capabilities of each individual; but clear signals of systematic distress among employees may be generated from quantifiable trends in physiological data, particularly when this is coupled with measurable performance information and medical observations about emotional state and behavior. The concept that psychological stress will be manifested in part by a significant change in some physical parameter provides the basis for developing useful early warning signals of incipient problems. Collecting and interpreting these data should be an appropriate function of the medical department. Many medical departments, often in conjunction with the internal counseling service, already have data of this type available, but there are inadequate channels established for using them effectively for organizational improvement.

The first objection to using health records in this manner is the possible violation of the confidential relation between the patient and his doctor. This need not be so. The signal output of initial interest is not how any particular individual is responding, but whether there are systematic trends. The message will, therefore, be statistical and numerical. (For example, one medical director suggested the simplest case of weighing existing medical files by department as a means of uncovering situations worth further investigation). By assigning the medical department responsibility for handling these inputs, protection of individuals can be assured.

Another potential barrier to using health assessment data may be concern about its cost. Outlays for medical services may already be at maximum justifiable levels in some organizations. Also, the health staff may not be eager to shift priorities from therapeutic to preventive activities. It seems probable that in many cases moving toward more involvement with management effectiveness will turn out to be an incremental cost and modest in amount. To begin with, substantial useful information may be generated simply by modifying the way in which medical data already being collected is processed and shared.

Although the experimental evidence for connection between measurable changes in physiological indices and stress as presented in the foregoing chapters seems convincing, there will be those who are reluctant to accept it. Even some who do accept it may feel that non-rational responses are so much influenced by early life experience that changing what is happening in the work place can have only minor effect in relieving stress reactions. No doubt, sorting out what the person brings to the job from what the job does to or for the person will be a necessary part of the techniques proposed to be developed.

Finally, there is the need for modifying the traditional role relationships and interdisciplinary communication channels within the organization so that effective return paths can be established for messages generated from health assessment data.

How can the potential value of this approach be tested? The first step should be to find organizations where top management and the medical staff consider such an effort sufficiently worthwhile to participate in some practical pilot studies to explore the possibilities. The Center for Occupational Mental Health has developed plans to act as a central, multidisciplinary agency for encouraging and assisting the conduct of each company's inital project.

Innovation of this type needs the enthusiastic support of the chief executive officer, as well as the expertise and commitment of the medical director and his staff, personnel directors, and senior line managers.

Next, a limited number of teams consisting of a line manager and a medical director from each organization will be brought together as a working group at the Center. With the assistance of qualified experts, this group will review various facets of the state-of-the-art and consider the needs and possible opportunities for applying it in their own organizations. Each team will then design a simple, feasible and measurable project involving a new style of cooperation between line managers and the health staff, which can be put into operation in each company during the ensuing six to twelve months. During that time the Center will serve as an information clearinghouse and furnish consultative support. The group will reconvene in about nine months to share experiences, critique results, and consider appropriate next steps. The outcome of the projects and the proceedings of the meetings will be published with suitable protection of any confidential information.

Initial projects will reflect careful assessment of the realities within the structure, climate and resources of each organization participating. They will range from regular meetings among an internal team of managers, medical staff, and personnel specialists for the purpose of clarifying and planning how they might best work together in a future project to a specific effort to review existing medical data, or, to begin collecting new data which can be tested immediately for its usefulness in management decision-making.

Among the results which are expected with some confidence, not the least is better understanding of the steps required to improve the communication network in each organization. There should also emerge more concrete appreciation of the probable expense involved in generating, processing, transmitting and utilizing the desired signals.

Over a three-year period it will be possible to develop enough sound data and experience to judge potential benefits and provide guidelines for techniques that might be used effectively by other organizations as well as the Center for Occupational Mental Health. No amount of theorizing can so quickly close the gap between what some now believe to be possible and what is actually possible in the day-to-day life of complex organizations. Given the urgency with which we must deal with general disaffection toward work in America at all levels, a pilot program under the auspices of the Center for Occupational Mental Health seems highly worthwhile.

CHAPTER 10

WORK PERFORMANCE AND OCCUPATIONAL STRESS*

BARRIE S. GREIFF

THE STUDY GROUP seriously considered the complexity of the issues. They recognized that, in spite of their expertise, the problems could only be outlined in the time available. They therefore asked themselves the following questions:

Correlation

Is there a correlation between occupational stress† and work performance, and is this correlation significant? In order to answer this

TABLE 10-I
HYPOTHETICAL RELATIONSHIPS BETWEEN PERFORMANCE
AND STRESS

*O.S.↑	**W.P.↑
O.S.↑	W.P.↓
O.S.↑	W.P. (remains the same)
O.S.↓	W.P.↑
O.S.↓	W.P.↓
O.S.↓	W.P. (remains the same)
O.S. (remains the same)	W.P.↑
O.S. (remains the same)	W.P.↓
O.S. (remains the same)	W.P. (remains the same)

*O.S. = Occupational Stress
**W.P. = Work Performance
↑ and ↓ Indicate increase or decrease
[Implied, also, is a feedback loop, where occupational stress effects work performance, and where work performance effects occupational stress]

---

*Participants in the subgroup which produced this report: John Anderson; Peter Brown; Arthur Danart; Royal Haskell, Jr., Ph.D.; George Kauer, Jr., M. D.; Jermyn F. McCahan, M. D.; W. Edward McGough, M. D.; Doris Minervini; David Ravella, M. D.; Porter Warren, M. D.; and Clinton Weiman, M. D.

†"Stress" is used to mean those environmental factors which stimulate unhealthy individual reactions.

question certain possibilities must be explored as exemplified in the following table of alternatives. Each in reality suggests a separate hypothesis and, possibly, separate research. (Table 10-I)

Work Performance

1. The definition of performance depends considerably upon those defining it—an important variable.

2. What criteria should be used to determine positive or negative work performance?

3. What is the type of organization involved? (e.g., profit versus non-profit, service versus industry, large versus small)

In any research various people must be consulted in order to determine a meaningful and comprehensive understanding of "total work performance." They include (a) the worker, (b) the supervisor, and (c) other significant individuals involved in determining performance. Intrapersonal, interpersonal, cultural, social, quantitative and economic criteria can be used as significant points for determining work performance; e.g., a worker may feel satisfied with his performance, (intrapersonal satisfaction), yet, as determined by other groups, his total productivity may be low, (managerial dissatisfaction). The problem can be stated thusly—

(a) Who determines the degree of work performance? (b) Using what criteria? (c) In what type of organization? (d) Involved in what type of culture? (e) and with what degree of expectation?

Occupational Stress

The effect of occupational stress upon a worker is dependent on:

1. The quality, intensity, and frequency of the stress agent or agents.

2. The susceptability of the worker. (As determined by past experiences, defense mechanisms, etc.)

Factors, in general, that promote a feeling of well-being among workers include the following:

1. The recognition of individual needs.

2. Feeling part of a decisional process.

3. The approval from supervisors.

4. Tangible evidence of success.

5. Security measures.

6. Open house opportunities.

7. Increased educational benefits.

The general factors included under occupational stress and which, in fact, might have a negative effect on work performance, include:

1. Boredom

2. Ambiguity

3. Role conflict

4. Lack of rewards

5. Loss of control of the worker and company, in terms of their objectives.

6. Reduced feedback, (work overload or underload), and interpersonal conflicts.

Summary

Whether there is a direct, significant correlation between occupational stress and work performance remains to be further investigated. Defining the terms stress, worker performance, and organization is necessary before any systematic investigation can be initiated. Finally, it should be noted, that when we talk about "the worker," in relationship to an organization, we must recognize that exclusive of the organization, the worker has had a past personal life, a family life, and lives in a culture with certain expectations and demands.

CHAPTER 11

OFF-THE-JOB STRESS AND OCCUPATIONAL BEHAVIOR*

ARTHUR B. SHOSTAK

AFTER CONSIDERABLE DISCUSSION of the many sources of off-the-job, personal pressures and problems that adversely effect workers—on the job as well as off—our subgroup decided to focus on four major areas of suggested reform.

To begin with, we urge constructive attention to gains possible from what we shall call, "aids to role expansion and enhancement." As an example, work enterprises might re-open the day care centers so popular during World War II. Rosie-the-Riveter brought the kids to the plant and they received pretty good care, and Rosie was better for having them there, better for being able to spend lunch time with them, and better for being able to take them home at the end of the day. Not only will reopening day care centers enable mothers to return to gainful employment, but professionals will be able to answer child-rearing questions and provide parents with special counseling during work or at lunch time. A good example of this is the KLH experiment, outside Boston, where mothers can bring their children while they receive job re-training.

We also urge corporations to pay more attention to the Masters-Johnson type of sex relations clinic that will soon be franchised across the country, hopefully with appropriate safeguards. If a large number of emotional problems, in fact, have their roots in inadequate sex education and sex relations, then corporate employees are paying a

---

*Participants in the subgroup: James L. Craig, M. D.; Paul Kurzman; Arlene Leonowicz, R. N.; Kerry Monick, M. D.; Robert Papp; Leon Rackow, M. D.; and Edward Seelye, M. D.

very heavy toll for their inadequacies. The new awareness of our needs in this area suggests that in the 1970's we are going to have a very genuine, perhaps somewhat unprecedented breakthrough in male-female relations linked to sex clinics.

Similarly, companies might imaginatively expand the responsibilities of their personnel specialists who have skills in counselling (recovery work, alcoholism, drugs, depression, etc.), and referrals. An example is described on page 17 of *OMH* (Volume 2, Number 3) (the Kennecott Copper "INSIGHT" program.) An extraordinary program, it is one of the best in personnel work and should serve as a model for industrial mental health.

Releasing professionals to help others on *and* off company property *and* time, as well as to work in tandem with a related labor union mental health program will not only accomplish much but will also offer an example of just what "role expansion and enhancement" (in this case, of the company personnel man, himself) is all about.

Second, we urge constructive programmatic attention to the gains possible from "aids to physical environment re-design." Specifically, we urge employers and unions alike to lobby vigorously in favor of anti-noise legislation, anti-litter bills, and pro-mass transit legislation. If mass transit isn't improved, then the employee arriving at work is going to be a Buchenwald-like survivor of the transit system and not the energetic guy we want showing up.

Zoning ordinances that provide for physical amenities in our buildings such as set-backs creating plazas, covered sidewalks, sidewalk cafes, and "street furniture" are to be championed, along with moves to close certain streets to cars altogether and to otherwise enhance the comfort and appeal of our physical world.

We recommend national legislation for vesting and pension protection. Workers are very anxious about conglomerates coming along and wiping out their pension protection. Today forty years of investment may add up to nothing. If we are interested in helping reduce stress, then we must see that employees receive a guaranteed retirement income.

We also are interested in government as employer of first resort. The 1946 Full Employment Act has never been enacted in this country. With the present 5½ percent unemployment rate it should

be activated and with some variants. For example, off-track-betting
in New York City guarantees that those employed who stay in a
heroin maintenance program will find posts even as the takers of
bets—presumably a sensitive post. We need more innovative programs as well as such things as quotas for minorities. Whatever the
Nixon administration has to say for Archie Bunker about abandon-
ment of the Philadelphia plan, in fact, they are going to keep pressing
with respect to minority quotas, which is probably as well.

Finally, we urge constructive attention to the gains possible from
"aids to adversary set scalebacks." In noncriminal matters we espouse
the cause of abortion-on-request and no-fault divorce, which California
has had for several years. In so-called "criminal matters, we favor
the withdrawal of criminal sanction from crimes with victims such as,
alcoholism, gambling, prostitution, pornography, homosexuality, and
narcotics use. Rather, we urge the substitution of money and genius
for rehabilitation and attractive alternatives. These private acts, when
conducted by or among consenting adults, are a private matter. If
we can reduce the number of adversary sets as listed, we can reduce
the stress these "sets" are causing.

Finally, we urge constructive attntion to the gains possible from
"aids to information enlargement." Some American businesses, for
example, are presently seeking to add a "human resources" section
to their annual report, measuring and assessing their ability to meet
workers' needs for job satisfaction, and to make careful and constructive calculations of morale an integral part of their accounting system.

Similarly, information about the health spa should be better publicized, so that more Americans might weigh this as a serious annual
option. Our European white-collar and blue-collar counterparts are
ahead of us in this regard. Many make regular use of the 3,000 or
so reconditioning centers throughout western Europe, including the
satellite nations, where companies and governments alike sponsor
stays that are prophylactic in orientation for up to a month for an
executive's family or a worker and his family.

Also, we urge that employment interviews at all levels be revised
to include an "extenuating circumstances" section. This would permit
both parties to the interview to sensitively explore critical information
that presently is not included, but nevertheless plays a large role in

shaping work performance (e.g., a preference for a relatively isolated work station; a fear of extended elevator stays; a desire to work with members of the same or a similar age, etc.).

In these four ways—or through new attention to role expansion, physical environment design, hostility scalebacks, and information enlargement—we can make substantial new gains against off-the-job stresses that flow over to depress on-the-job performance. While there was a consensus in the subgroup that these four suggestions were of major importance, unanimity did not prevail completely; the chairman must accept responsibility for this report.

CHAPTER 12

OCCUPATIONAL "STRESS"—A MISNOMER

ALAN McLEAN

A T THE TIME the decision was made to focus this conference on occupational stress I was relatively sure what the term "stress" meant. My concept, a clinical one, was not far afield from many of my colleagues. I felt comfortable using it in work with patients reacting to "stressful" situations in their work environments. For years I considered an occupational "stress" or "stressor" (and I loosely interchanged the terms) any work-related factor which produced a maladaptive response. Most of these symptomatic responses could be labeled with a broad array of diagnoses—generally to be found in the diagnostic manual of the American Psychiatric Association. But I also included adverse effects on work performance and on interpersonal relationships. Conceptually, I was aware of Selye's definition and use of the term and of the more exact stipulations of others.

During the meetings it became clear that both the term "stress," and its relationship to adaptation at work, is used in such widely varying ways as to suggest we abandon the word entirely. I am reminded of the old computer concept "garbage in, garbage out." There is a tremendous amount of garbage in the literature—a great deal of fuzzy thinking. Imprecise concepts of occupational stress continued to be manifest at this conference. We did, however, develop a sense of understanding of the many points of view. Most speakers defined their terms with great clarity and numerous examples.

People from different disciplines tend to speak only to those in the same field with a bland disregard of significant work of writers from a different frame of reference. Lennart Levi's volume *Society, Stress and Disease* (Levi, 1971) illustrates the problem with forty-four separate papers by authors who tend to use the term "stress" in different ways. Sidney Cobb put it succinctly in his review of the Levi book, "The time has come for us to speak in specific terms . . . We should begin to speak of role conflict and role overload, and role ambiguity in the same way we speak of pneumococus types I, II and III. Perhaps the concept of stress will remain useful in the same general sense as the concept of infection. But, by the same token that no specialist in infectious diseases would be satisfied to diagnose a patient as simply having an infection, we should no longer be satisfied to say that a patient is suffering from social stress, or that a particular environment is stressful." I agree we must be more specific although I am not yet comfortable with the concepts of overload, ambiguity and conflict.

DEFINITIONS

Many at these meetings used the engineering analogy—the concept that stress is a force applied that induces strain or deformation in that to which it is applied. For engineers this is a clear, precise and useful concept. External loading producing internal stress leading to internal strain. The external loading may cause overloading producing overstress in turn producing yielding (irreversible strain). Such yielding may not yet prevent functioning though it may lead in turn to breakdown or rupture. Frequent repetition of stress and strain may produce fatigue from which we then produce rupture without a phase of overloading.

Some of those who follow the engineers, borrowing their concepts and applying them to human behavior, tend to define stress as a stimulus situation, the properties of which are response-inferred. Stress is defined as an extreme or noxious stimulus which generally results in certain physiological change, behavioral change, perceptual change, etc. It produces both overt and intrapsychic coping efforts (Kahn and Quinn, 1970). The stress is defined by the responses of a whole class of organisms rather than by responses of a particular individual.

Others following the engineers, verbalized a prevailing clinical viewpoint. Wright, for instance, said:

> The term stress implies an inherent capacity to resist or stand up to a defined amount of strain. Increasing the loading on a girder beyond

STRESS OR STRAIN IN ENGINEERING TERMS

LOADING (external) ----- ------▶STRESS (internal) ------▶STRAIN (internal)

    may cause

OVERLOADING -----▶OVERSTRESS ---▶YIELDING   (irreversible   strain)

                     which may not yet prevent

                           function
                             ↓

                  BREAKDOWN or RUPTURE

Figure 12-1. Frequent repetition of stress and strain may produce FATIGUE which can then produce RUPTURE without a phase of OVERLOADING (ICOMH, 1971).

a certain point will inevitably bend it. Individuals too have their breaking point, and if this is exceeded they will also bend or break. In practice, various, largely unconscious defense mechanisms are thrown into play which include psychosomatic reactions. These tend to lead the individual to opt out of the stressful situation and can be regarded as a direct result of the interplay between the individual and his immediate environment.

Clearly this is a complicated situation in which the individual's reaction is the final common pathway between his training, attributes and personality and the various specific and cultural pressures that the environment—work, home and community—are putting on him. (Wright, 1967).

Wright, and many of the rest of us, have found it convenient to express this as the personality/environment equation: a dynamic, constantly changing, and reactive relationship between the individual, acting as a totality, in his whole environment.

Wright expresses concern that this approach leads to the total neglect of "the beneficial aspects of stress" or indeed the inevitability of some stress or challenge in the maintenance of biological life and effectiveness (yet another commonly held notion). Indeed Taylor agrees,

We think of stress as something that is harmful or deleterious to performance and, in general, we wish to eliminate it. However, we would do well to note that stress is often accompanied by, and indeed may be a necessary part of, the process of change and growth. It is a paradox that many of our actions are designed to reduce stress, but consider what would result if we achieved all our goals and eliminated all of stress. Without some degree of stress, there would be no productivity and almost certainly there would be boredom. When the stress factor is nonexistent we are dead." (Taylor 1967).

And again we see examples of the differing use of the term. Even when using the engineering analogy some tend to see "stress" as a constructive force; others view it only as productive of strain. Some see individual reactions; others the reaction of a whole class.

The examples cited above were drawn from both practitioners and researchers; from those principally concerned with patient care and through working at a conceptual level. Obviously the engineering concept has drawn to it a sizeable body of proponents who have found the model useful. Few, however, use it in the same way; I am sure that an engineer would be appalled by the distortions he sees as clinical and behavioral scientists borrow his physical theories. In fact it is so misapplied that we now have several sociological, psychological and physiological meanings for the term which are diverse and inconsistent.

PHYSIOLOGICAL CONCEPTS OF STRESS

When we shift our discussion to the work of those concerned with physiological, hormonal and biochemical reactions to environmental stimuli, one most often thinks of the work of Hans Selye. Here, in contrast to the uses of the term "stress" cited above, stress is the *reaction* to a stressor or rather a variety of non-specific stressors. In Selye's terms it is associated with a very specific syndrome—the "general adaptation syndrome." Since first described by Selye in 1936 a great number of biochemical and structural changes of previously unknown etiology have been traced to non-specific stress. Among these, general attention has been given by clinicians to alterations in blood lipid and electrolyte patterns as well as the characteristic rise in the serum level of acute phase reactants that occur during the first stage of the general adaptation syndrome. Considerable work in the analysis of hormonal mediation of stress reactions has also, of course, been described.

As Selye pointed out at Levi's Stockholm conference,

> Biologic stress is the non-specific response of the body to any demand made upon it. Stress cannot be defined as the consequence of excessive demands. In the living body stress is inseparably linked with the response to any kind of demand, normal or excessive. In the living organism excessive use only accelerates and intensifies the development of stress manifestations, and thereby facilitates their detection.
>
> The syndrome of stress (the general adaptation syndrome) is quite specific in its manifestations but non-specific in its causation.
>
> Finally, we must clearly distinguish between stressors (the causative agent) and stress (the non-specific response). Cold is not a stress but a stressor. The use of the term 'cold stress' is justified only when it is meant to designate the combination of manisfestations produced by cold as a specific agent and its non-specific stressor effect. (Selye, 1971).

A specific "stressor" has been defined as a change in the internal or external environment of such magnitude (qualitative or quantitative) that it requires from the organism more than the usual adaptation and defense reactions to maintain its life and/or homeostasis. A stressor is therefore different from a stimulus (which means any change in the environment) because of its intensity, and the difference between a stimulus and a stressor is often quantitative rather than qualitative. Moreover, it depends also on the sensitivity of the organism at a certain moment, whether a stimulus or event is a stressor or not. Another difference is that we usually apply the term stressor only to an environmental change in relation to the total organism, and not when studying isolated organs or tissues.

Stressors (like stimuli) can be manifold in nature: physical, chemical, viral, bacteriological, biological and interhuman.

> Interhuman conflicts (actual or anticipated) are the most common stressors. If acted out in the form of violence, an interhuman conflict may result in bodily damage. In most cases, however, interhuman conflicts are acted out verbally or symbolically; and in this case they threaten or upset primarily the homeostasis of the central nervous system, and through this that of the total organism. (Groen, 1971).

An interesting demonstration by Levi related to the clinical comments of Wright and Taylor. He has shown that stress thus defined increases during *pleasant* emotional reactions such as laughter and sexual arousal. Moreover, many individuals deliberately seek stress in the form of adventure, excitement, challenge. Clearly, stress may very well occur and, consequently, the rate of wear and tear on the organism may increase even when the individual experience is a pleasant emotional reaction. The immediate effects are also desirable from the viewpoint of society such as in the case of great devotion to work (Levi, 1971).

Clearly then, Groen, Selye and Levi and a great number of their colleagues also find a legitimate use of the term "stress." Those aligned with this viewpoint such as Dr. Stewart Wolf feel that the term "stress" might well be dispensed with *in the psychological context*. They tend to retain it for their own use (Stress [Selye]). Wolf points out that what one customarily talks about *psychologically* is the behavior of a person and his bodily equipment in response to the environment—that is, briefly, *adaptation* (Wolf, 1971).

The Stockholm conference clearly underlined the differences between stress (Selye)—the stereotyped physiological reaction pattern— and psychological stress—that is "distress."

INTERVENING VARIABLES

Between a stressor and the individual reaction to it are a host of variables—seemingly infinite in number. They apply whether we discuss psychological, psychosocial or physiological reactions:

1. Biochemical individuality may explain some of the great differences between individuals in reaction to identical stimuli. Rather consistent inter-individual differences in hormonal response to psychosocial stressors is known, as are the wide ranges within which each person's hormone level fluctuates under ordinary circumstances. There may indeed be inborn errors of hormonal metabolism, including a block in synthesis and a defect in transport and disposal sytems.

2. Early life experiences.

3. Psychological set. Both physiological and psychological reactions to various stimuli clearly depend upon the so-called "neuroendocrine tone" of the individual at any particular time.

4. Cultural factors often determine the stressor potential of various psychosocial stimuli. The differences in culturally assigned roles, for example, vary widely and in turn both help define stressors and determine individual and group reaction.

5. Both conscious and unconscious mechanisms of defense come into play. If one is prepared to cope with a specific situation or stressor, the reaction is much less than if one is not. The consequences of the same threat are not understandable without reference to the kinds of coping activity that the individual engages in and the kinds of personality that lead him to use one type of coping process or another.

CONCLUSIONS

After listening to the distinguished participants in this conference it is my inclinaton to agree with Lazarus (1966, 1971) that "stress" had best be considered a general rubric for a large collection of related problems rather than a single narrow concept. If we were to use the word stress to refer, say, only to Selye's adaptation syndrome, then its meaning would be restricted entirely to physiological stress reaction to stressor and in fact to an even narrower concept: namely the non-specific responsive tissues. In so doing we would eliminate stress-induced adaptational problems at the level of behavioral mal-adaptation including serious psychopathology and many varieties of subclinical problems of uncomfortable individuals who are not necessarily suffering from a tissue disease.

Stress, then, I must conclude is neither stimulus, response, nor in-tervening variable, but rather a collective term for an area of study It might even be well to allow the term "stress"—and in particular, occupational stress—to be extended as Lazarus suggests (Lazarus, 1971). It could easily refer to a very broad class of problems dif-ferentiated from other problem areas because it deals with any demands which tax the sytem whatever it is—a physiological system, a social system, or a psychological system; the response of that system must be included. Each discipline must further develop its own terms to refer to the *specific* concepts of its own analysis. The pre-sentations of each researcher and clinician *must* define his frame of reference *and* his terms in such manner as to be intelligible to those in his professional audience unfamiliar with what has gone on before in his discipline.

Clarity, definition and exposure; cross-pollenation and acceptance of these differing concepts as deeply held values is urgently necessary if we are to further our task.

BIBLIOGRAPHY

Groen, Joannes J.: Social change and psychosomatic disease. In Levi, Lennart (Ed.): *Society, Stress and Disease.* New York, Oxford University Press, 1971.

International Committee on Occupational Mental Health: *Stress in Industry.* The Committee, London, 1971.

Kahn, Robert L., and Quinn, Robert P.: Role stress: A framework for analysis. In McLean, Alan (Ed.): *Mental Health and Work Organizations,* Chicago, Rand, 1970.

Lazarus, R. S., and Opton, E. M., Jr.: The study of psychological stress: A summary of theoretical formulations and experimental findings. In Spielberger, C. D. (Ed.): *Anxiety and Behavior.* New York, 1966.

Lazarus, R. S.: Environmental planning in the context of stress and adaptation. In Levi, Lennart (Ed.): *Society, Stress and Disease.* New York, Oxford U Pr, 1971.

Levi, Lennart: *Society, Stress and Disease.* New York, Oxford University Press, 1971.

Selye, Hans: The evolution of the stress concept—stress and cardiovascular disease. In Levi, Lennart (Ed.): *Society, Stress and Disease,* New York, Oxford University Press, 1971.

Taylor, Graham C.: Executive stress: In McLean, Alan (Ed.): *To Work Is Human.* New York, Macmillan, 1967.

Wolf, Stewart: Discussion—Page 453. In Levi, Lennart (Ed.): *Society, Stress and Disease.* New York, Oxford University Press, 1971.

Wright, H. Beric: Executive stress in Great Britain. In McLean, Alan (Ed.): *To Work Is Human.* New York, Macmillan, 1967.

INDEX